# PLASMA PHYSICS

## An Introduction to Statistical Physics of Charged Particles



Setsuo Ichimaru
University of Tokyo

The Benjamin/Cummings Publishing Company, Inc.
Menlo Park, California • Reading, Massachusetts
Don Mills, Ontario • Wokingham, U.K. • Amsterdam • Sydney
Singapore • Tokyo • Mexico • Bogota • Santiago • San Juan

Sponsoring Editor: Richard W. Mixter
Production Editor: Karen Gulliver



Library of Congress Cataloging-in-Publication Data

Ichimaru, Setsuo, 1935-
Plasma physics.

Based on: Purazuma no butsuri. Tokyo; Sangyo Tosho, 1981.
Includes index.
1. Plasma (Ionized gases) I1 Ichimaru, Setsuo, 1935- . Purazuma no butsuri. II. Title.
QC718.I24 1985 530.4'4 85-18488
ISBN 0-8053-8754-4

ABCDEFGHIJ-MA-898765

The Benjamin/Cummings Publishing Company, Inc.
2727 Sand Hill Road
Menlo Park, California 94025

## Editor's Foreword

Everyone concerned with the teaching of physics at the advanced undergraduate or graduate level is aware of the continuing need for a modernization and reorganization of the basic course material. Despite the existence today of many good textbooks in these areas, there is always an appreciable time-lag in the incorporation of new viewpoints and techniques which result from the most recent developments in physics research. Typically these changes in concepts and material take place first in the personal lecture notes of some of those who teach graduate courses. Eventually, printed notes may appear, and some fraction of such notes evolve into textbooks or monographs. But much of this fresh material remains available only to a very limited audience, to the detriment of all. This series aims at filling this gap in the literature of physics by presenting occasional volumes with a contemporary approach to the classical topics of physics at the advanced undergraduate and graduate level. Clarity and soundess of treatment will, we hope, mark these volumes, as well as the freshness of the approach.

Another area in which the series hopes to make a contribution is by presenting useful supplementing material of well-defined scope. This may take the form of a survey of relevant mathematical principles, or a collection of reprints of basic papers in a field. Here the aim is to provide the instructor with added flexibility through the use of supplements at relatively low cost.

The scope of both the lecture notes and supplements is somewhat different from the "Frontiers in Physics" series. In spite of wide variations from institution to institution as to what comprises the basic graduate course program, there is a widely accepted group of "bread and butter" courses that deal with the classic topics in physics. These include: Mathematical methods of physics, electromagnetic theory, advanced dynamics, quantum mechanics, statistical mechanics, and frequently nuclear physics, solid-state physics, and plasma physics. It is chiefly these areas that will be covered by the present series. The listing is perhaps best described as including all advanced undergraduate and graduate courses which are at a level below seminar courses dealing entirely with current research topics.

Finally, because the series represents a continuing experiment on the part of the editors and the publisher, suggestions from interested readers as to format, contributors, and contributions will be most welcome.

DAVID PINES

# Preface

In nature, or in the laboratory setting, we encounter various physical systems containing numerous charged particles such as electrons and ions. These mobile, charged particles interact with each other via long-range electromagnetic forces, and thereby endow interesting mediumlike properties to the system. Such systems are generally called plasmas. In this book we take up plasma phenomena, investigate their salient features, and study how these features may be accounted for in theoretical terms.

A number of alternative approaches appear to exist in the study of the physics of plasmas. One can, for example, analyze equilibrium and nonequilibrium plasma phenomena systematically from the point of view of the statistical mechanics of charged particles. These are "elegant" approaches, so to speak, where one attempts to deduce the fundamental properties of the plasma, starting with the first principles in physics. This would offer a most efficient approach if one had a prior knowledge of the nature of the plasma problem to be solved.

When we deal with the actual plasma phenomena, however, such an efficient method does not always apply. In many cases, the raw observational or experimental data contain the consequences of various physical effects. In order to extract and elucidate the effect germane to plasma physics, it is sometimes necessary to follow rather contrived and laborious processes of learning directly from what the nature indicates to us, attempting to interpret them from the fundamental principles, returning to the observational indications, revising the interpretation, and so forth. In this sense, the physics of plasma may belong to "empirical sciences" or "applied physics."

In this book, by way of emphasizing such empirical aspects, we take note of the experimental or observational appearances of various plasma phenomena in early stages of chapters, and attempt to elucidate, in terms of those experimental results, the characteristic features of plasmas as charged many-particle systems. As for the examples of actual plasmas; we take up the conduction electrons and holes in metals and semimetals, ionized gases in the laboratory, magnetic fusion experiments, the free-electron laser, the ionosphere, dense matter in stellar interior, the pulsars, and X-ray stars. Through investigations of these various plasma systems we intend to explore what the outstanding problems are in the world of plasmas, whose parameters range over a wide domain.

Since the book is written with experimental aspects placed early in the presentation, there may be instances where the logical flows of expositions are somewhat impeded. In some cases derivations of basic theoretical devices necessary for the explanation are deferred to later stages of chapters or to appendixes. I wish to advise the reader, at the time of first reading, not to be troubled by such irregularities and simply to keep on reading through the entire volume. At each stage of the topical developments, a best effort has been expended to give an account of the relevant physical processes involved.

The primary purpose of this book will have been achieved if it is able to convey the sense of what a fascinating subject plasma physics is, and thereby to induce the reader to a further study of the statistical physics of charged particles revealed in the rich varieties of plasma phenomena in nature or in the laboratory. At a second reading of the book, the reader may undertake a detailed study of those various plasma processes with the aid of the appendixes and the cited references, which would then serve as a useful guide to a further advanced learning.

The author has taught the subjects contained in this book to undergraduate students in physics at the University of Tokyo for the past several years; the material presented here is an outgrowth of those lectures. The presentation of this book is self-contained and should be read without difficulty by those who have adequate preparation in classical mechanics, electricity and magnetism, elementary quantum mechanics, and statistics. In addition to ordinary calculus, we use such mathematical tools as complex variables and Fourier analysis; some of the appendixes are devoted to exposition of those mathematical techniques. References are cited mainly for the purpose of assisting the reader in finding relevant discussions extending and supplementing the material presented here; this is especially the case for Chapters 10 and 12, where somewhat advanced topics are treated. No claims are intended to priority or superiority of a given reference cited.

Chapters 1-4, 6, 7, 9, 11, and 12 and Appendixes 1, 2, 5 and 7 derive from the translation of a plasma-physics book written originally in Japanese by the present author and published by Sangyo Tosho Publishing Co., Tokyo, Japan. Substantial revisions have been made in the course of the translation. The remaining chapters and appendixes are newly added to this English version.

I should like to express my sincere gratitude to Professor Kazuhiko Nishijima, who suggested I write the original Japanese version, and to Professor David Pines, who encouraged me to do this English version. Past and present associates with my laboratory at the University of Tokyo, namely, Dr. Kenichi Utsumi, Dr. Kazutami Tago, Dr. Hiroshi Iyetomi, Mr. Shinichi Mitake, and Mr. Shigenori Tanaka, gave me useful comments on the manuscript and provided assistance in the preparation of the drawings. Mrs. Mariko Ohashi and Mrs. Keiko Uchikoshi carefully typed the entire manuscript and extended able assistance at various phases of this work. I am indebted to Mr. Takehiko Ezura of Sangyo Tosho Publishing Co., for his editorial assistance in the publication of the original Japanese version, and to Mr. Richard W. Mixter of the Benjamin/Cummings Publishing Co., for this publication. Finally, I wish to thank my wife, Tomoko Ichimaru, who undertook the original English translation, and gave me considerable help in the completion of this project.

This work was supported, in part, through Grants-in-Aid for Scientific Research, Nos. 56380001 and 59380001, provided by the Japanese Ministry of Education, Science, and Culture.

Tokyo, Japan
July 1985

Setsuo Ichimaru

**CONTENTS**

# Chapter 1

# Plasmas in Nature

Atoms and molecules, building blocks of ordinary substances, consist of nuclei and electrons. An **electron** has a negative unit charge ($e = 4.803 \times 10^{-10}$ cgs esu). **Ions** produced by removing some or all of the orbital electrons from atoms or molecules carry positive charges. A system containing many of such relatively mobile charged particles is called a **plasma.** This statement cannot be regarded as a definition of the plasma in a strict sense because it contains ambiguous terms such as "many" and "relatively mobile." Leaving the problem of contriving a rigorous definition to the interested reader, let us look at a number of examples of plasmas occurring in nature.

## 1.1 Examples of Plasmas

When "plasma" is mentioned, many readers may be reminded that it has something to do with "nuclear fusion." This is a scheme of devising a new energy source by exploiting nuclear energy released in fusion processes between light nuclei of hydrogen isotopes such as deuterium and tritium. Since atomic nuclei are positively charged, they repel each other electrostatically. To induce nuclear fusion reactions effectively by overcoming such repulsive forces, those nuclei have to collide vigorously with each other. The minimum conditions for net production of energy are estimated to be a confinement of dense, high-temperature plasma of $10^{14}$ - $10^{15}$ $cm^{-3}$ and $\sim 10^8$ K for more than a second. Stable confinement and heating of a plasma are essential problems in realization of the controlled release of nuclear fusion energy.

Plasmas are produced in the laboratory by **electric discharge.** When a strong electric field is applied to an ordinary gas, the charged particles in it are accelerated.* The charged particles accelerated to high energies collide with other atoms or molecules, and thereby ionize those neutral particles by removing orbital electrons. Such a process of producing ions through dissociation of electrons is generally called **ionization.** Those ions and electrons produced by collisional ionization are then accelerated by the electric field, and then ionize other neutral particles by collisions; ionization processes thus proceed like an avalanche. This is a phenomenon of the electric discharge. The **ionized gas** produced by electric discharge is a typical example of plasmas; it is found commonly in our daily life such as in the fluorescent lamp, neon tube, and in arc welding.

The electron densities and temperatures of the plasmas produced by electric discharge vary widely depending on the condition of the discharge. For example, in the **glow discharge** at the gas pressure of $\sim 1$ Torr, one usually finds the electron density of $10^9$ - $10^{11}$ $cm^{-3}$ and the electron temperature of $\sim 10^4$ K.

---

* In a neutral gas such as air, electrons and ions do exist, though infinitesimal in quantity, due to ionizing action of e.g., cosmic rays.

Without using an electric discharge, a plasma state may be attained by simply raising the temperature of a neutral gas. In these circumstances, thermal energy of the particles is utilized for impact ionization and hence **thermal ionization.** Since the binding energy of an electron in a neutral molecule takes on the order of a few electron volts, the temperature* must correspondingly exceed $10^4$ - $10^5$ K.

As a plasma is a collection of mobile charged particles, it is an electrically conductive fluid; when an external electromagnetic field is applied to a plasma an electric current is induced in it. The electric current so induced acts to change the motion of the plasma through coupling with the electromagnetic field. For example, an electric current flowing in a direction perpendicular to the magnetic field exerts a force in the direction perpendicular to both of those fields (the motor action). When the motion of a plasma is changed in an electromagnetic field, a flow pattern of the electric current in the plasma is now altered analogously due to the generator action. One thus observes an intimately related action between the plasma motion and the electromagnetic field as a consequence of the plasma being an electrically conducting fluid. The **magnetohydrodynamics** deals with the problems in this field; it directly concerns the macroscopic behavior of the plasma (e.g., Spitzer 1962; Alfvén and Fälthammer 1963; Krall and Trivelpiece 1973).

Similarily, **metals** or **semiconductors** contain relatively mobile, free charged particles in them. For instance, one finds conduction electrons at a density of approximately $10^{23}$ $cm^{-3}$ in a metal; a semiconductor contains mobile electrons and

---

* 1 eV corresponds to 11,605 K, through conversion with the use of the Boltzmann constant.

holes, though their densities are substantially smaller than that. Those electrically conductive carriers can be regarded as a sort of plasma. In particular the system of conduction electrons in metal is called a **quantum plasma**; because of its high density, quantum-mechanical interference effects between neighboring electrons become essential.

Above the surface of the earth, layers of ionized gas exist at altitudes of 70 - 500 km over the stratosphere; this is the **ionosphere**, useful in radio communication. The electron density is largest in the upper part, at altitudes 200 - 500 km (F-layer). Its value is approximately $10^6$ cm$^{-3}$, and the electron temperature is in the vicinity of 2000 K, although those values vary depending on day or night, seasons, and solar activities.

As the influence of the solar activities on the ionosphere may suggest, the sun itself is an important source of plasmas. A flow of plasma, the **solar wind**, is ejected from the sun into the interplanetary space. Wind velocity near the earth is about 250 - 700 km/s; the number density of plasma particles is 1 - 10 cm$^{-3}$, and the electron temperature is $\sim 10^5$ K. The solar wind collides with the magnetic field of the earth at about 10 - 13 $R_E$ ($R_E$ = 6400 km: the radius of the earth), confines the magnetic field in it, and thereby produces the boundary surface of the domain called the **magnetosphere.**

As for the sun, it has a region containing solar atmosphere called the **chromosphere** (a few thousand km in depth). It is a plasma of an average electron density $10^{10}$ - $10^{11}$ cm$^{-3}$, and temperature $\sim 10^4$ K. Above the chromosphere there extends a large region of **corona** (observed in familiar photographs of the solar eclipses). Its plasma parameters such as the electron density depend to a great extent on the distance from the solar surface. At $10^5$ km, the density and temperature are estimated approximately as $10^8$ cm$^{-3}$, and $10^6$ K.

The **stellar interior**, as well as the stellar atmosphere, is in a plasma state. The mass density and temperature of the solar interior are approximately $10^2$ g/cm$^3$, and $10^7$ K. The internal densities of those stars in advanced stages of stellar evolution take on much larger values. For a **white dwarf** (a degenerate star), one of the final stages of stellar evolution, its average density reaches as high as $10^5$ - $10^8$ g/cm$^3$, corresponding to the electron density $10^{28}$ - $10^{31}$ cm$^{-3}$.

Finally we take up the **interstellar space** containing those stars; although dilute, it also makes an example of plasmas. Typical electron densities range $10^{-2}$ - $10^0$ cm$^{-3}$, and the temperature $\sim 10^4$ K. Furthermore we find various phenomena related to plasmas in those conspicuous astrophysical phenomena such as, the **supernova remnants, pulsars, X-ray stars, and active galactic nuclei,** including the **quasars.**

As a summary of this section, we show in Fig. 1.1 a schematic presentation of the parameter domains for various plasmas occurring in nature.

## 1.2 Plasma Parameters

As we have overviewed in the previous section, the plasmas found in nature or in the laboratory are characterized by the temperature and density parameters over a wide range of variation. We thus find it convenient to introduce a dimensionless coupling constant, which characterizes strength of the particle interactions in those charged-particle systems with widely varying parameters.

For simplicity, we consider a spatially homogeneous **one-component plasma** (OCP). It is a system consisting of a single species of charged particles embedded in a uniform background of neutralizing charges. Such an OCP is a substantially

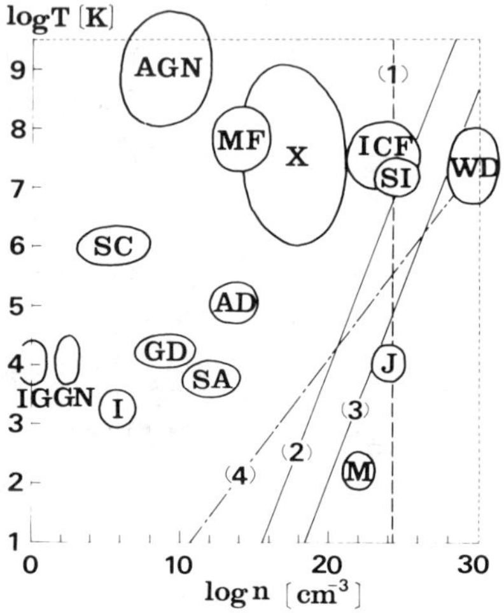


**Figure 1.1.** Typical values of the electron density n and the temperature T for various plasmas: interstellar gas (IG), gaseous nebula (GN), ionosphere (I), glow discharge (GD), solar atmosphere (SA), arc discharge (AD), solar corona (SC), active galactic nucleus (AGN), magnetic fusion (MF), X-ray star (X), inertial confinement fusion (ICF), solar interior (SI), metal (M), Jovian interior (J), and white dwarf (WD). The lines represent: (1) $r_s = 1$, (2) $k_BT/E_F = 10$, (3) $k_BT/E_F = 0.1$, and (4) $\Gamma' \equiv (e^2/k_BT)(4\pi n/3)^{1/3} = 1$.

idealized model for real plasmas; some plasmas in nature, as we shall see, do indeed satisfy the conditions for such idealization.

Let us define a **coupling constant** of the plasma as

(Coulomb coupling constant)

$$= \frac{\text{(average Coulomb-interaction energy)}}{\text{(average kinetic energy)}} . \qquad (1.1)$$

We are thus concerned with the strength of coupling due to Coulomb interaction. Those plasmas with values of the coupling constant much smaller than unity may be called **weakly coupled plasmas**; those with the coupling constant around or greater than unity, **strongly coupled plasmas.**

We first estimate the average kinetic energy appearing in the denominator of Eq. (1.1). When the charged-particle system under consideration obeys the classical statistics, the velocity distribution of the individual particles is given by the Maxwellian,

$$f(\vec{\mathbf{v}}) = \left(\frac{M}{2\pi k_B T}\right)^{3/2} \exp\left(-\frac{Mv^2}{2k_B T}\right) . \tag{1.2}$$

Here M denotes the mass of a particle, $k_B = 1.3807 \times 10^{-16}$ erg/deg is the Boltzmann constant, and T is the temperature. The average value of the kinetic energy per particle is then calculated as

$$\left\langle \frac{1}{2} Mv^2 \right\rangle = \int_0^\infty dv 4\pi v^2 \left(\frac{1}{2} Mv^2\right) f(\vec{\mathbf{v}}) = \frac{3}{2} k_B T . \tag{1.3}$$

We are concerned with an order-of-magnitude estimation of the coupling constant, Eq. (1.1), and so we take the average kinetic energy per particle to be $k_BT$, ignoring the forefactor 3/2 in Eq. (1.3).

For a high-density electron system such as the conduction electrons in metal, one uses the **Fermi energy**

$$E_F = \frac{\hbar^2}{2m} (3\pi^2 n)^{2/3} \tag{1.4}$$

instead of the classical evaluation mentioned above. Here $\hbar = 1.0546 \times 10^{-27}$ erg · s is the Planck constant divided by $2\pi$, m $= 9.1095 \times 10^{-28}$ g is the electron mass, and n denotes the number density of the electrons. Since the electron is a fermion, a quantum state is occupied by a single electron at most (Pauli exclusion principle). The Fermi energy corresponds to the energy of the final n-th electron when the quantum states in a unit volume are occupied by n electrons successively from the lowest energy level. Since Eq. (1.4) is an increasing function of n, $E_F >> k_BT$ may be realized in a high-density electron system. In these circumstances, the quantum effects as fermions become much more important than the effects of thermal motion represented by $k_BT$; it is relevant to use the Fermi energy as an estimate of the kinetic energy. In such a **degenerate electron system,** the effects arising from the Fermi statistics thus play an essential role.

The Coulomb energy per particle of an OCP containing n particles with electric charge Ze in a unit volume may also be estimated as $(Ze)^2/a$, where

$$a = (4\pi n/3)^{-1/3} \tag{1.5}$$

is the radius of a sphere with the characteristic volume 1/n; this radius is usually referred to as the **ion-sphere radius** or the **Wigner-Seitz radius.**

The coupling constant, Eq. (1.1), for the degenerate electron system is

$$\frac{e^2/a}{E_F} = 0.543\ r_s \tag{1.6}$$

where

$$r_s = \left(\frac{3}{4\pi n}\right)^{1/3} \frac{me^2}{\hbar^2} \tag{1.7}$$

is the Wigner-Seitz radius for electrons measured in units of the **Bohr radius,** $a_B = \hbar^2/me^2 = 5.292 \times 10^{-9}$ cm. As Eq. (1.6) indicates $r_s$ is equivalent to the Coulomb coupling constant for the degenerate electron system; $r_s = 1$ corresponds to $n = 1.6 \times 10^{24}$ cm$^{-3}$.

The number density n of the conduction electrons in metal may be estimated as

$$n \simeq (\text{valence}) \times \frac{(\text{mass density})}{(\text{mass number})} \times \frac{1}{M_n} , \qquad (1.8)$$

where $M_n \simeq 1.67 \times 10^{-24}$ g is the mass of a nucleon. The values of $r_s$ estimated on the basis of Eq. (1.8) for various metals at room temperatures are 2.1(Al), 2.7(Mg), 3.3(Li), 4.0(Na), 5.0(K), and 5.8(Cs); hence $r_s = 2 - 6$ for ordinary metals. The coupling constant thus takes on a value close to, or greater than unity. On the other hand, we rewrite Eq. (1.4) in terms of $r_s$ to find

$$E_F = 50.11\ r_s^{-2}[\text{eV}] = 5.82 \times 10^5\ r_s^{-2}[\text{K}] . \qquad (1.9)$$

For $r_s < 6$, $E_F \gg k_BT$ holds at room temperatures (~ 300 K). The electron system in metal is a typical example of the degenerate, strongly coupled plasmas.

The coupling constant for a plasma obeying the classical statistics is

$$\Gamma = \frac{(Ze)^2/a}{k_BT} = 2.69 \times 10^{-5} Z^2 \left[\frac{n}{10^{12}\ \text{cm}^{-3}}\right]^{1/3} \left[\frac{T}{10^6\ \text{K}}\right]^{-1} . \qquad (1.10)$$

This expression indicates that for a plasma with $Z \simeq 1$ and $T \simeq 10^6$ K, the density n must become as high as ~ $10^{26}$ cm$^{-3}$ to make $\Gamma \simeq 1$.

Most of the classical plasmas that we ordinarily encounter, however, are characterized by $\Gamma \ll 1$. For example, we may assume $n \simeq 10^{11}$ $cm^{-3}$, $T \simeq 10^4$ K for a gaseous-discharge plasma, $n \simeq 10^{16}$ $cm^{-3}$, $T \simeq 10^8$ K for a plasma in a controlled thermonuclear experiment, and $n \simeq 10^6$ $cm^{-3}$, $T \simeq 10^6$ K for a plasma in the solar corona. For those plasmas, respectively, we find $\Gamma \simeq 10^{-3}$, $10^{-5}$, and $10^{-7}$. They are thus weakly coupled plasmas; their thermodynamic properties are analogous to those of an ideal gas. According to the theoretical calculation (Abe 1959), the internal energy density U of a weakly coupled plasma is expressed as

$$\frac{U}{nk_BT} = \frac{3}{2} - \frac{\sqrt{3}}{2}\Gamma^{3/2} - 3\Gamma^3\left[\frac{3}{8}\ln 3\Gamma + \frac{\gamma}{2} - \frac{1}{3}\right] \qquad (1.11)$$

where $\gamma = 0.57721$ ..... is Euler's constant.

A most typical example of a strongly coupled classical plasma may be seen in the system of ions inside a highly evolved star. The interior of such a star is in a compressed, high-density state. The Fermi energy of the electron system takes on a value much greater than the binding energy of an electron around an atomic nucleus; all the atoms are thus in ionized state (**pressure ionization**). The electron system constitutes a weakly coupled ($r_s \ll 1$) degenerate plasma with an immensely large Fermi energy ($E_F \simeq mc^2$). It makes an ideal neutralizing background of negative charges to the ion system; since $r_s \ll 1$, the polarizability of the background may be negligible. Those atomic nuclei stripped of the electrons form an ion plasma obeying the classical statistics; their de Broglie wavelengths are much smaller on the average than the interparticle spacing, that is, $\hbar(Mk_BT)^{-1/2} \ll a$, where M is the ionic mass. In the interior of a highly evolved star, the coupling constant $\Gamma$ for such an ion plasma is usually greater than unity; in a white dwarf one estimates that $\Gamma = 10 - 200$ (e.g., Schatzman 1958; Van Horn 1971).

The thermodynamic properties of a strongly coupled OCP has been determined accurately by means of computer simulation of the Monte Carlo type (Brush, Sahlin, and Teller 1966). According to the recent investigation (Slattery, Doolen and DeWitt 1980, 1982), the internal energy density U is expressed as

$$\frac{U}{nk_BT} = - 0.897744\Gamma + 0.95043\Gamma^{1/4}$$

$$+ 0.68513 + 0.18956\Gamma^{-1/4}, \quad (1 \leq \Gamma \leq 170) \; . \qquad (1.12)$$

As an example of a strongly coupled plasma closer to the earth, we may take the Jovian interior, which consists mostly of hydrogen and a small fraction of helium. Its electron system is characterized by $r_s = 0.6 - 1$; its ion system, $\Gamma = 20 - 50$. The Jovian interior thus appears to offer a complex system of strongly coupled classical plasma immersed in a polarizable background of degenerate electrons. In addition, the ion plasma may have to be looked upon as a **binary ion mixture** of hydrogen and helium.

As an example of a strongly coupled plasma in the laboratory, one may think of plasmas produced in the laser implosion experiments (Brueckner and Jorna 1974), whose densities will eventually reach values as high as $\Gamma = 1 - 200$. It has been reported that $\Gamma = 1 - 5$ has been realized in transient plasmas produced in explosive shock tubes (Ivanov, Mintsev, Fortov, and Dremin 1976). In addition, as the purest example of strongly coupled classical OCP realized in the laboratory, we have the two-dimensional system of electrons trapped in the surface states of liquid helium (Cole 1974; Grimes 1978).

The physical properties of a strongly coupled plasma differ in an essential way from those of a weakly coupled plasma. As the coupling constant is further increased, it is said that a strongly coupled plasma undergoes crystallization. Those interesting properties of the strongly coupled plasma are currently being investigated from various directions (Baus and Hansen 1980; Ichimaru 1982).

Chapter 2

# Plasma Properties Due to the Coulomb Interaction

The extent of Coulomb coupling depends on the temperature and density of the plasma. The Coulomb coupling constant, Eq. (1.1), gives a measure on the strength of such a coupling. In a plasma with a large coupling constant, the charged particles are strongly correlated with each other; the plasma behaves as a kind of "liquid." This does not mean, however, that one can ignore the Coulomb interaction in a weakly coupled plasma, which would behave like a "gas." A cooperative process of weak coupling effects arising from participation of a huge number of particles may produce a sizable, observable effect. It is in this sense that the Coulomb interaction plays an essential part in determination of the system properties even for a weakly coupled plasma. The cooperative effect stems from the long-range nature of the Coulomb potential, which varies inverse proportion to the distance. In general many other charged particles may be involved within the force range of a given charged particle. The physics of plasmas is thus intimately connected to the **many-body problem.** In this chapter, we shall consider salient properties of plasmas arising from the many-body effects.

## 2.1 Scattering Cross-Section for Momentum Transfer

The effect of particle interaction in a plasma may be appreciated first by a consideration of scattering event between two particles. In the process of scattering, parts of the momentum and energy carried by one particle are transferred to the other particle. The **cross-sections for momentum and energy transfer** describe the efficiencies by which those conserved physical quantities are transferred from one particle to the other in a collisional event.

To calculate the cross-section for momentum transfer, one multiplies the differential cross-section of scattering by the rate of momentum change per collision, and then integrates the product over all the scattering angles. Let us consider two charged particles, $Z_1e$ and $Z_2e$, in a plasma; their reduced mass and relative velocity are $\mu$ and $v$ in the center-of-mass system. Following a standard method in the classical mechanics, one finds the relation between the scattering angle $\theta$ and the impact parameter b (Fig. 2.1) as

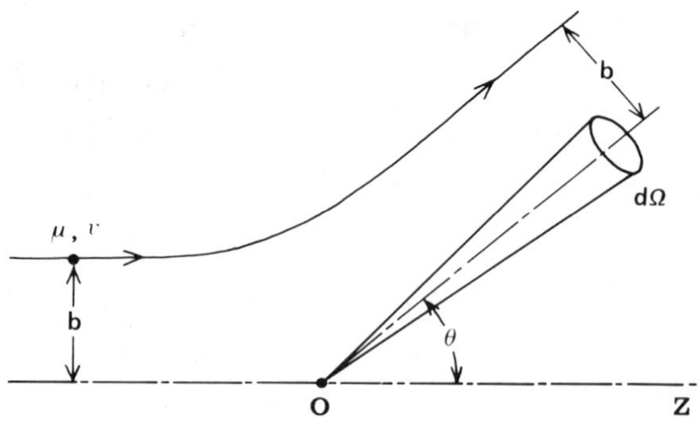


**Figure 2.1.** Coulomb scattering

$$\cot \frac{\theta}{2} = \frac{b\mu v^2}{Z_1 Z_2 e^2} . \tag{2.1}$$

The differential cross-section dQ for scattering into an infinitesimal solid angle $d\Omega$ is then given by

$$\frac{dQ}{d\Omega} = \left[\frac{Z_1 Z_2 e^2}{2\mu v^2 \sin^2(\theta/2)}\right]^2 . \tag{2.2}$$

This is the formula for **Rutherford scattering.**

The increment of momentum in the direction of OZ axis in Fig. 2.1 is $\Delta p = \mu v(1 - \cos\theta)$. Since the scattering is symmetric with respect to the OZ axis, the increments of momentum in the other two directions vanish on the average. Consequently, the cross-section $Q_m$ for the momentum transfer due to Coulomb scattering is calculated as

$$Q_m = \int_{\theta_{min}}^{\pi} (1 - \cos\theta)\left[\frac{Z_1 Z_2 e^2}{2\mu v^2 \sin^2(\theta/2)}\right]^2 2\pi \sin\theta \, d\theta$$

$$= 4\pi \left[\frac{Z_1 Z_2 e^2}{\mu v^2}\right]^2 \ell n\left[\frac{1}{\sin(\theta_{min}/2)}\right] . \tag{2.3}$$

We have deliberately written the lower limit of the integral in Eq. (2.3) as $\theta_{min}$; if we let $\theta_{min}$ approach zero, $Q_m$ would diverge logarithmically. According to Eq. (2.1), a small scattering angle $\theta$ corresponds to a large impact parameter b. The divergence in Eq. (2.3) thus originates from cummulative effects of large-distance scattering, reflecting the long-range nature of the Coulomb interaction.

When $Z_1e$ and $Z_2e$ are separated at a large distance in a plasma, many other plasma particles may be found between them; hence the effective interaction between $Z_1e$ and $Z_2e$ is altered because of the presence of those other plasma particles. In other words, in the calculation of the cross-section such as Eq. (2.3), one must take account of the effective potential including the many-body effects of other charged particles. This is an important problem because the cross-section for the momentum transfer is directly related to the evaluation of the electric conductivity and viscosity.

## 2.2 Debye Screening

Consider a weakly coupled classical plasma ($\Gamma \ll 1$) with the electric charge $Ze$, temperature $T$[K], and number density $n$[cm$^{-3}$]. In this section we take up the problem of determining the effective electrostatic potential around an external electric charge $Z_0e$ introduced in the plasma.

We choose the position of the external charge at the origin of the coordinates, which are expressed by the radial vector $\vec{r}$. In the absence of the plasma, the electrostatic potential is given simply by the Coulomb law,

$$\phi_0(\vec{r}) = \frac{Z_0 e}{r} . \tag{2.4}$$

In the presence of a plasma, such an electrostatic potential induces a deviation of the space-charge density from the average. If $Z$ and $Z_0$ are of the same sign, plasma particles are repelled from the external charge; the probability of their approaching in its vicinity would be reduced. If they are of the opposite sign, they would attract each other and approach closer. The effective potential $\phi(\vec{r})$ stemming from introduction of the external charge is calculated as a summation of Eq. (2.4) due directly to the external charge and the potential due to the induced charge density. The Poisson equation then is

$$\nabla^2\phi(\vec{r}) = -4\pi Z_0 e\delta(\vec{r}) - 4\pi Ze\langle\rho(\vec{r})\rangle , \tag{2.5}$$

where $Ze\langle\rho(\vec{r})\rangle$ refers to the induced charge density and $\delta(\vec{r})$ is the three-dimensional delta-function representing the density distribution of a point particle located at the origin, $\vec{r} = 0$ (cf. Appendix 1).

The deviation $\langle\rho(\vec{r})\rangle$ from the average plasma density induced by the effective potential $\phi(\vec{r})$ is calculated according to the Boltzmann distribution as

$$\langle\rho(\vec{r})\rangle = n \exp\left[-\frac{Ze\phi(\vec{r})}{k_B T}\right] - n . \tag{2.6}$$

The argument, $-Ze\phi(\vec{r})/k_B T$, of the exponential function represents the ratio of the potential energy $Ze\phi(\vec{r})$ per particle to the average kinetic energy $k_B T$. When $\Gamma \ll 1$, this ratio may likewise be regarded as sufficiently small, so that we may expand Eq. (2.6) as*

$$\langle\rho(\vec{r})\rangle \simeq -\frac{Zen\phi(\vec{r})}{k_B T} . \tag{2.7}$$

Substitution of Eq. (2.7) in Eq. (2.5) yields a differential equation to determine the effective potential $\phi(\vec{r})$,

$$-\nabla^2\phi(\vec{r}) + \frac{4\pi n(Ze)^2}{k_B T}\phi(\vec{r}) = 4\pi Z_0 e\delta(\vec{r}) . \tag{2.8}$$

A way of solving this equation is to use the method of

---

* In the vicinity of $r = 0$, this expansion is inaccurate even with $\Gamma \ll 1$. In the calculation of the effective potential, however, this difficulty does not show up explicitly. For an example of such a difficulty, see Sec. 6.4.

Fourier transformation. We summarize the essence of Fourier analysis in Appendix 2, where the solution to Eq. (2.8) is also described, with the result:

$$\phi(\vec{\mathbf{r}}) = \frac{Z_0 e}{r} \exp\left(-\frac{r}{\lambda_D}\right) . \tag{2.9}$$

The parameter

$$\lambda_D \equiv \left[\frac{k_B T}{4\pi n (Ze)^2}\right]^{1/2} \tag{2.10}$$

introduced here is called the **Debye length**; it is an important characteristic length for a weakly coupled classical plasma. Numerically, it takes an expression

$$\lambda_D = 6.90 \times 10^{-3}\ Z^{-1}\left[\frac{n}{10^{12}\ \mathrm{cm}^{-3}}\right]^{-1/2}\left[\frac{T}{10^6\ \mathrm{K}}\right]^{1/2}\ (\mathrm{cm}) . \tag{2.11}$$

The meaning of Eq. (2.9) is clear by comparison with Eq. (2.4). For $r < \lambda_D$, the exponential factor in Eq. (2.9) takes on a value close to unity; the effective potential is virtually identical to Eq. (2.4). For $r > \lambda_D$, on the contrary, $\phi(\vec{\mathbf{r}})$ takes on a value much smaller than $\phi_0(\vec{\mathbf{r}})$ due to the exponential factor; for $r \gg \lambda_D$, we may take $\phi(\vec{\mathbf{r}}) \simeq 0$. In other words, when the distance is substantially greater than the Debye length, the electrostatic potential of the external charge $Z_0 e$ is effectively screened by the space charge induced in the plasma. This is called **Debye screening** (Debye and Hückel 1923).

Let us consider the number of plasma particles involved in such an act of Debye screening. For this purpose we calculate the average number of particles $N_D$--to be called the **Debye number**--contained in a sphere with a radius of the Debye length:

$$N_D \equiv \frac{4\pi}{3} \lambda_D{}^3 n = (3\Gamma)^{-3/2}$$

$$= 1.38 \times 10^6 \ Z^{-3} \left[\frac{n}{10^{12} \ \mathrm{cm}^{-3}}\right]^{-1/2} \left[\frac{T}{10^6 \ \mathrm{K}}\right]^{3/2} . \tag{2.12}$$

As we find here, the Debye number takes on an extremely large value for a weakly coupled plasma ($\Gamma \ll 1$). On the other hand, the total electric charge necessary for screening of the electrostatic potential produced by the external charge $Z_0e$ is only $-Z_0e$; this amounts to a deviation of $|Z_0/Z|$ from the average in terms of the number of particles. Ordinarily we may assume $|Z_0| \simeq |Z|$, so that the number is close to unity. Consequently, for realization of Debye screening in a weakly coupled plasma, so many ($\sim N_D$) charged particles act cooperatively by an extremely small amount each ($\sim N_D{}^{-1}$), to erase the effect of the external charge field. It is in this sense that the Debye screening makes an illuminating example of the cooperative phenomena brought about by the long-range Coulomb interaction in the plasma.

## 2.3 The Coulomb Logarithm

On the basis of the Debye-screening concept, let us review the calculation of the scattering cross-section for momentum transfer described in Section 2.1. We take both $Z_1e$ and $Z_2e$ as particles in the plasma and thus replace them with $Ze$. Since the range of the effective potential in the plasma is on the order of the Debye length, we may determine the lower limit $\theta_{min}$ of the $\theta$ integration in Eq. (2.2) as a solution to Eq. (2.1) with $b = \lambda_D$. We also use the Maxwellian-averaged value,

$$\langle \frac{\mu}{2} v^2 \rangle = \frac{3}{2} k_B T , \tag{2.13}$$

for the relative velocity on its right-hand side, to find

$$\cot\left(\frac{\theta_{min}}{2}\right) \simeq 3\,\frac{\lambda_D k_B T}{(Ze)^2} = 9N_D \ . \tag{2.14}$$

For $N_D \gg 1$, one has $\theta_{min} \ll 1$, so that $\theta_{min}/2 = 1/9N_D$. Consequently Eq. (2.3) may be evaluated as

$$Q_m = 4\pi\left(\frac{Z^2e^2}{\mu v^2}\right)^2 \ln(9N_D) \ . \tag{2.15}$$

The factor $\ln(9N_D)$ appearing here is an example of the **Coulomb logarithm.** It reflects the logarithmic divergence in the scattering cross-section associated with the Coulomb interaction. Use of Eq. (2.12) yields

$$\ln(9N_D) = 16.3 + 1.15\log\left\{Z^{-6}\left[\frac{n}{10^{12}\ \mathrm{cm}^{-3}}\right]^{-1}\left[\frac{T}{10^6\ \mathrm{K}}\right]^3\right\} \ . \tag{2.16}$$

Hence we find in many cases that, not only $N_D \gg 1$, but also $\ln(9N_D) \gg 1$ holds true. In these circumstances, estimation of $\theta_{min}$ does not call for a high accuracy; it suffices to maintain a logarithmic accuracy as seen numerically in Eq. (2.16).

## 2.4 The Ion-Sphere Model

Thus far we have considered a problem of the effective potential in a weakly coupled plasma and derived the concept of Debye screening. As we move into the strong coupling domain ($\Gamma \geq 1$),the Debye number becomes smaller than unity; the concept of Debye screening as a cooperative phenomenon is no longer applicable. The probability of finding other charged particles in a sphere with the radius $\lambda_D$ almost vanishes. A charged particle creates a sizable domain around itself where no other charged particles are likely to be found, a sort of a territorial domain of its own influence which may be looked upon as a Coulomb hole.

To understand salient features of such a strongly coupled plasma, it is instructive to introduce the ion-sphere model (Salpeter 1954). It is equivalent to the Wigner-Seitz sphere used in the solid-state physica (e.g., Pines 1963). As Fig. 2.2 illustrates, one considers a charged particle Ze and a surrounding neutralizing charge sphere, whose total electric charge is just to cancel the charge Ze. This sphere thus represents the territorial domain of influence for the charge Ze. Its radius is a of Eq. (1.5) and the charge density, $-3Ze/4\pi a^3$.

The ion sphere thus consists of a single ion and its surrounding negative-charge sphere. The electrostatic energy $E_{IS}$ associated with the ion sphere is then calculated in the following way: First the electrostatic potential produced by the negative-charge sphere at $r(\leq a)$ is

$$\phi_{IS}(r) = -\frac{3Ze}{4\pi a^3}\int_{r'\leq a} d\vec{r}'\frac{1}{|\vec{r}-\vec{r}'|} = -\frac{3Ze}{2a} + \frac{Ze}{2a}\left(\frac{r}{a}\right)^2 . \qquad (2.17)$$

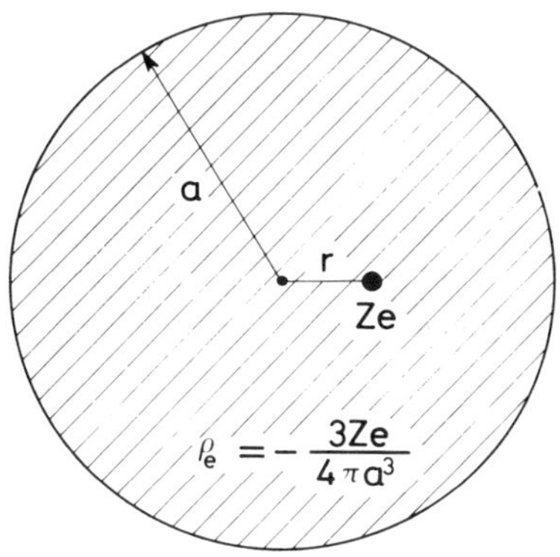


**Figure 2.2.** Ion-sphere model. $\rho_e$ refers to the average charge density in the sphere.

The electrostatic energy of the negative-charge sphere itself is then

$$-\frac{1}{2}\int_{r\leq a} d\vec{r}\,\frac{3Ze}{4\pi a^3}\,\phi_{IS}(r) = \frac{3}{5}\frac{(Ze)^2}{a} \quad . \tag{2.18}$$

Summation of this energy and $Ze\phi_{IS}(r)$ yields

$$\frac{E_{IS}}{k_B T} = -\,0.9\,\Gamma + 0.5\,\Gamma\left(\frac{r}{a}\right)^2 \quad . \tag{2.19}$$

The first term on the right-hand side of Eq. (2.19) represents the electrostatic energy when the ion Ze is located at the center of the ion sphere. The second term being proportional to $r^2$ induces a motion of the harmonic-oscillator type to the ion. The average energy of a harmonic oscillator including the kinetic energy is $k_B T$ per a degree of freedom. The density of internal energy calculated on the basis of the ion-sphere model is thus

$$\frac{U_{IS}}{nk_B T} = -\,0.9\,\Gamma + 3 \quad . \tag{2.20}$$

This result, in fact, takes on a value close to the computer simulation result such as Eq. (1.12) in the domain of $\Gamma \gg 1$ (Ichimaru 1982).

The ion-sphere model has played an important part in elucidating the properties of the strongly coupled plasma; its validity has been demonstrated through various examples (Ichimaru 1982). Here we consider how the ion-sphere model may be applied to the calculation of the cross-section for momentum transfer, Eq. (2.3).

According to the ion-sphere model of Fig. 2.2, the range of the electrostatic potential of the charged particle Ze is confined within a distance $\alpha_1 a$, where $\alpha_1$ is a correction factor of order unity accounting for uncertainty involved in a strict enforcement of the ion-sphere model. Putting $Z_1 = Z_2 = Z$ in Eq. (2.1), we determine the scattering angle $\theta_{min}$ corresponding to the impact parameter at the distance $\alpha_1 a$,

$$\cot\left(\frac{\theta_{min}}{2}\right) = \alpha_1 \frac{a\mu v^2}{(Ze)^2} \quad . \tag{2.21}$$

Equation (2.3) then becomes

$$Q_m = 2\pi\left(\frac{Z^2e^2}{\mu v^2}\right)^2 \ln\left\{1 + \left(\frac{\alpha_1 a\mu v^2}{Z^2e^2}\right)^2\right\} \quad . \tag{2.22}$$

For a strongly coupled plasma ($\Gamma \gtrsim 1$), we may take $\alpha_1 a\mu v^2/(Ze)^2 < 1$ for a majority of particles; retaining only the first term after expanding the logarithmic term in Eq. (2.22), we obtain

$$Q_m \simeq 2\pi(\alpha_1 a)^2 \quad . \tag{2.23}$$

As we would anticipate from the construction of the ion-sphere model, the scattering cross-section is proportional to the cross-section $\pi a^2$ of the ion sphere.

Equation (2.23) thus takes a form easy to interpret qualitatively. Quantitatively, however, it depends strongly on the estimation of the undetermined parameter $\alpha_1$. A theoretical determination of $\alpha_1$ is a complex problem that calls for a detailed knowledge of interparticle correlations in the strongly coupled plasma (Baus and Hansen 1980; Ichimaru 1982). Later in Chapter 10, we shall study in some detail the correlation properties of strongly coupled plasmas and apply the results to calculation of the enhancement factor of the thermonuclear reaction rate in dense stellar matter.

## 2.5 Plasma Oscillation

We have been concerned with the static properties of the plasma, such as the screening effect, which do not involve a time-dependent behavior of density variations in the plasma. The long-range Coulomb interaction produces not only such a static cooperative effect but also the plasma oscillation, which may be regarded as a dynamic cooperative phenomenon. To treat this problem, we begin with consideration of the equation of motion describing the temporal behavior of density fluctuations in the plasma (e.g., Pines 1964).

Let $\vec{r}_j(t)$ be the spatial coordinates of j-th particle at time t of the n charged particles contained in a unit volume. In terms of the three-dimensional delta-function introduced in Eq. (2.5), the density distribution arising from the j-th particle is $\delta[\vec{r} - \vec{r}_j(t)]$. Summing this for all the n particles constituting the plasma, we express the particle-density distribution as

$$\rho(\vec{r}, t) = \sum_{j=1}^{n} \delta[\vec{r} - \vec{r}_j(t)] \quad . \tag{2.24}$$

Using the periodic boundary conditions appropriate to the unit volume (cf. Appendix 2), we calculate the Fourier component of the density fluctuations with wave vector $\vec{k}$. Since the three-dimensional delta-function has the property

$$\int_{\text{volume containing } \vec{r}_0} d\vec{r}\,\delta(\vec{r} - \vec{r}_0)y(\vec{r}) = y(\vec{r}_0) \quad , \tag{2.25}$$

we obtain from Eq. (2.24)

$$\rho_{\vec{k}}(t) = \int_{\text{unit volume}} d\vec{r}\, \rho(\vec{r}, t)\exp(-i\vec{k} \cdot \vec{r})$$

$$= \sum_{j=1}^{n} \exp[-i\vec{k} \cdot \vec{r}_j(t)] \quad . \tag{2.26}$$

To derive an equation of motion, we differentiate this density fluctuation twice with respect to time. As $\dot{\vec{r}}_j = \vec{v}_j$ represents the velocity of the j-th particle, we find

$$\ddot{\rho}_{\vec{k}} = - \sum_j \{(\vec{k} \cdot \vec{v}_j)^2 + i\vec{k} \cdot \dot{\vec{v}}_j\}\exp(-i\vec{k} \cdot \vec{r}_j) \quad . \tag{2.27}$$

The quantity $\dot{\vec{v}}_j$ representing the acceleration of the j-th particle can be calculated from the force exerted upon this particle by all the other charged particles and the neutralizing background charge,

$$\dot{\vec{v}}_j = - \frac{1}{M} \frac{\partial}{\partial \vec{r}_j} \sum_{\ell(\neq j)} \frac{(Ze)^2}{|\vec{r}_j - \vec{r}_\ell|} \tag{2.28}$$

$$+ \text{ (acceleration by the background charge).}$$

Here, M denotes the mass of a particle and $\partial/\partial\vec{r}_j$ means a vectorial differential operation--gradient--with respect to the $\vec{r}_j$ coordinates.

As explained in Appendix 2, the Fourier expansion (L = 1) of the Coulomb potential is given by

$$\frac{Ze}{r} = \sum_{\vec{k}} \frac{4\pi Ze}{k^2} \exp(i\vec{k} \cdot \vec{r}) \quad .$$

We apply this expression in Eq. (2.28), where we note that

the $\vec{k} = 0$ component cancels exactly with the acceleration due to the neutralizing charge; we thus obtain

$$\dot{\vec{v}}_j = -i\,\frac{4\pi(Ze)^2}{M}\sum_{\vec{q}}{}'\,\frac{\vec{q}}{q^2}\,\rho_{\vec{q}}\exp(i\vec{q}\cdot\vec{r}_j) \quad . \tag{2.29}$$

Here the prime at the shoulder of the summation sign $\sum$ implies that the $\vec{q} = 0$ term is to be excluded in the process of the summation.

Substituting Eq. (2.29) in Eq. (2.27), we find

$$\ddot{\rho}_{\vec{k}} = -\sum_j(\vec{k}\cdot\vec{v}_j)^2\exp(-i\vec{k}\cdot\vec{r}_j) - \frac{4\pi(Ze)^2}{M}\sum_{\vec{q}}{}'\,\frac{\vec{k}\cdot\vec{q}}{q^2}\,\rho_{\vec{k}-\vec{q}}\,\rho_{\vec{q}} \quad . \tag{2.30}$$

The first term on the right-hand side represents the effect of translational motion of the individual particles; the second term stems from the Coulomb interaction. We particularly single out the $\vec{q} = \vec{k}$ term in the interaction term; since $\rho_0 = n$, Eq. (2.30) is rewritten as

$$\ddot{\rho}_{\vec{k}} + \omega_p^2\rho_{\vec{k}} = -\sum_j(\vec{k}\cdot\vec{v}_j)^2\exp(-i\vec{k}\cdot\vec{r}_j) - \frac{4\pi(Ze)^2}{M}\sum_{\vec{q}(\neq\vec{k})}{}'\,\frac{\vec{k}\cdot\vec{q}}{q^2}\,\rho_{\vec{k}-\vec{q}}\,\rho_{\vec{q}} \quad . \tag{2.31}$$

Here we have introduced another important parameter

$$\omega_p \equiv \left[\frac{4\pi n(Ze)^2}{M}\right]^{1/2} \tag{2.32}$$

called the **plasma frequency**; for an electron plasma it is computed as

$$\omega_p = 5.64 \times 10^{10} \left[\frac{n}{10^{12}\ \mathrm{cm}^{-3}}\right]^{1/2} \quad (\mathrm{s}^{-1}) \ . \tag{2.33}$$

If the right-hand side of Eq. (2.31) is neglected altogether, the temporal variation of the density fluctuation with wave vector $\vec{k}$ obeys an equation of motion for the harmonic oscillator whose characteristic frequency is given by $\omega_p$ (Tonks and Langmuir 1929 a, b; Langmuir 1929). As we see in Eq. (2.26), $\rho_{\vec{k}}$ contains the coordinates of many charged particles; nevertheless, we here observe a possibility of such a quantity exhibiting an orderly oscillatory behavior under the action of the long-range Coulomb forces, an appearance of the **collective motion.** In the next section we investigate the conditions under which the density fluctuations behave collectively in a plasma.

## 2.6 Collective Motion and Individual-Particlelike Behavior

Carrying out a Maxwellian average of the velocity variables contained in the first term on the right-hand side of Eq. (2.31), we obtain

$$\sum_j (\vec{k}\cdot\vec{v}_j)^2 \exp(-i\vec{k}\cdot\vec{r}_j) \simeq \sum_j \exp(-i\vec{k}\cdot\vec{r}_j)\int d\vec{v}(\vec{k}\cdot\vec{v})^2 f(\vec{v})$$

$$= k^2\left(\frac{k_B T}{M}\right)\rho_{\vec{k}} \ . \tag{2.34}$$

The second term, on the other hand, represents a nonlinear term consisting of the products of density fluctuations. On the complex-number plane as shown in Fig. 2.3, the Fourier component of density fluctuations is expressed as a summation of unit vectors over the total number of particles; the phase angle of each vector is given by $-\vec{k}\cdot\vec{r}_j$ . For a uniform

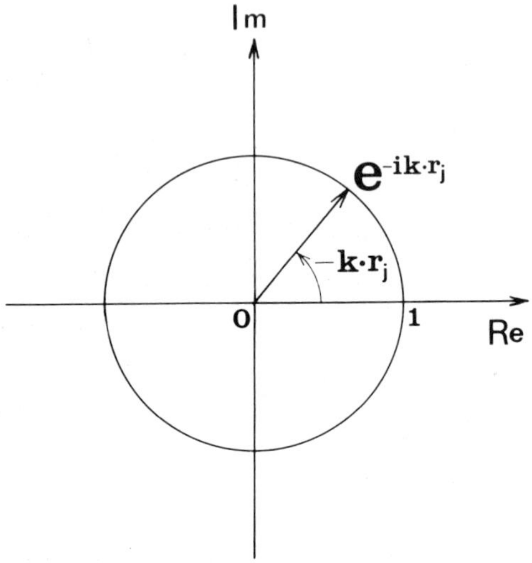


**Figure 2.3.** Representation of density fluctuation on a complex-number plane.

plasma, such phase angles distribute randomly as long as $\vec{k} \neq 0$; hence the expectation value $\langle \rho_{\vec{k}} \rangle = 0$.

Bohm and Pines (1951, 1953; Pines and Bohm 1952) considered a possibility that an analogous situation may become applicable to the second term on the right-hand side consisting of products of density fluctuations. They thereby introduced an approximation, called the **random-phase approximation** (RPA), in which those product terms are ignored. Even if the phase $-\vec{k} \cdot \vec{r}_j$ is distributed randomly, however, the effect arising from the product of fluctuations $\rho_{\vec{k}-\vec{q}}\,\rho_{\vec{q}}$ cannot generally be neglected. Setting the motivation of the original nomenclature aside, we may reinterpret the RPA as an approximation applicable to the cases of weak fluctuations whereby the equation of motion may be linearized with respect to the fluctuations. The RPA thus provides a fairly accurate description of a weakly coupled plasma near thermodynamic equilibrium. In these circumstances we may ignore the second term on the right-hand side of Eq. (2.31), as they are nonlinear terms of fluctuations.

Let us thus adopt this RPA. The condition that the density fluctuations behave collectively and produce plasma oscillations may be derived from a comparison between the second term on the left-hand side of Eq. (2.31) and Eq. (2.34) as

$$k^2 \ll k_D{}^2 \; . \tag{2.35}$$

The parameter

$$k_D = \left[\frac{4\pi n (Ze)^2}{k_B T}\right]^{1/2} \tag{2.36}$$

is the inverse of the Debye length defined in Eq. (2.10); we shall call it **Debye wave number.**

In the long-wavelength domain satisfying Eq. (2.35), the density fluctuations in the plasma behave collectively and oscillate at a frequency in the vicinity of $\omega_p$. In this domain the frequency spectrum of fluctuations may look as shown schematically in Fig. 2.4.

In the short-wavelength domain such that $k^2 \gg k_D{}^2$. On the other hand, the first term on the right-hand side of Eq. (2.31) is the major contribution and the equation of motion becomes

$$\ddot{\rho}_{\vec{k}} = - \sum_j (\vec{k} \cdot \vec{v}_j)^2 \exp(-i\vec{k} \cdot \vec{r}_j) \; . \tag{2.37}$$

This is nothing but the equation of motion for fluctuations in the system where the Coulomb interaction is neglected altogether and individual particles are assumed to move on rectilinear trajectories without changing velocities $\vec{v}_j$. Consequently, the fluctuations in this case may be looked upon as a superposition of the random motion of discrete individual particles. The frequency spectrum of fluctuations now exhibits

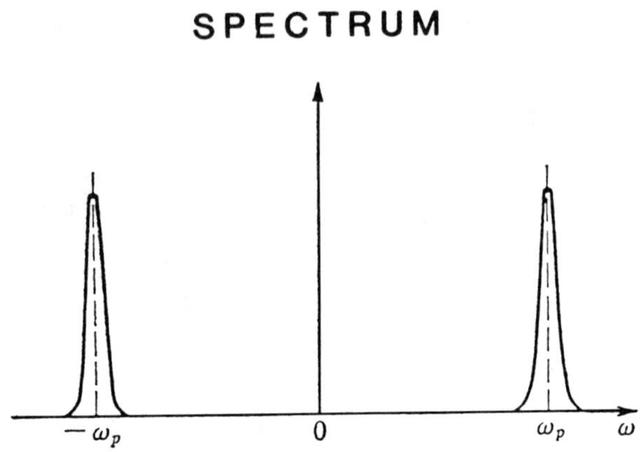


**Figure 2.4.** Frequency spectrum of density fluctuations in the long-wavelength domain--collective motion.

a characteristic thermal spread as shown in Fig. 2.5.

The individual-particlelike spectrum of Fig. 2.5. is a characteristic feature of a neutral-particle system of an ideal-gas type. In a plasma, such an individual-particlelike

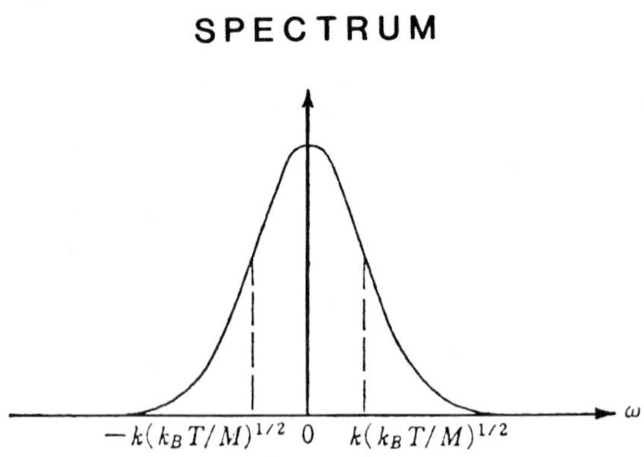


**Figure 2.5.** Frequency spectrum of density fluctuations in the short-wavelength domain--individual-particlelike fluctuations.

behavior in the short-wavelength domain coexists with the collective motion, that is, the plasma oscillation in the long-wavelength domain; the latter represents the principal effect of the Coulomb interaction. Such a coexistence of the collective and individual-particlelike behaviors adds a variety of interesting features to the properties of the plasma, as we shall see in the following sections.

## 2.7 Wave-Particle Interaction

As mentioned in the previous section, the plasma wave as a collective motion and the random individual-particle motion coexist in a plasma. Those waves and particles interact with each other and produce a variety of new effects. In this section, we take up a couple of the simplest among those effects. Some of the nonlinear effects will be considered later in Chapter 9.

As a consequence of the wave-particle interaction, the characteristic frequency of the density fluctuation at $\vec{k} \neq 0$ deviates from $\omega_p$, resulting in an additional dispersion term proportional to $k^2$. Plasma waves with different wave numbers thus propagate with different phase velocities $\omega/k$ and different group velocities $\partial\omega/\partial k$. A possible existence of such a dispersion term may be inferred from the first term on the right-hand side of Eq. (2.31). According to a calculation on the basis of the RPA (see Appendix 5 for derivation), the frequency $\omega_k$ of the plasma oscillation is given by

$$\omega_k{}^2 = \omega_p{}^2 + 3\langle v^2\rangle k^2 \tag{2.38}$$

where $\langle v^2\rangle$ is the mean square of the particle velocity in the direction of the wave vector $\vec{k}$. With the Maxwellian distribution (1.2), Eq. (2.38) becomes

$$\omega_k^2 = \omega_p^2 + 3\left(\frac{k_B T}{M}\right)k^2 . \tag{2.39}$$

Equation (2.38) or (2.39) is referred to as the **Bohm-Gross dispersion relation** (Bohm and Gross 1949 a, b).

The amplitude of the plasma wave decays through resonant interaction with the particles. This mechanism may be understood in the following way: We assume a plasma wave with wave number k and frequency $\omega_k$ having an amplitude sufficiently small so that a linearized treatment is applicable. Figure 2.6 shows the potential variation associated with the space-charge distribution in the frame co-moving with the wave. Those charged particles with velocities almost the same as the phase velocity $\omega_k/k$ of the wave interact resonantly with the wave; they continue to perceive almost the same wave potential. The condition for the wave-particle resonance is thus expressed as

$$\omega = \vec{k} \cdot \vec{v} . \tag{2.40}$$

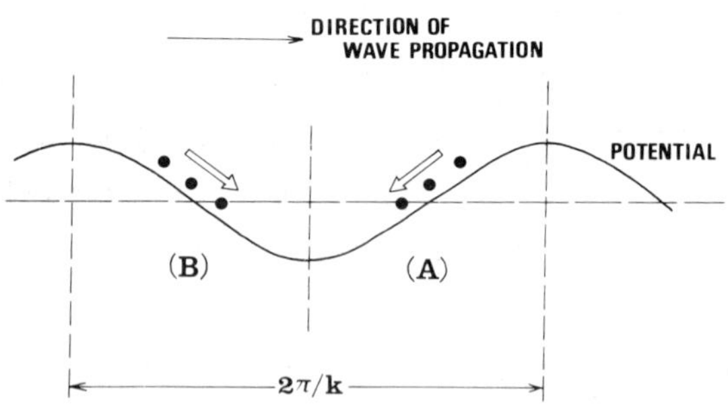


**Figure 2.6.** Resonant interaction between plasma waves and particles.

Those among the resonant particles whose velocities are slightly faster than the phase velocity of the wave will act to push the wave in the domain (A) of Fig. 2.6; they are thus decelerated by the wave. Conversely, those particles with velocities slightly slower than the phase velocity of the wave, situated in the domain (B) of the Fig. 2.6, are accelerated by the wave. The former particles thus act to amplify the wave, while the latter induce decay of the wave. Consequently, whether the wave amplitude would actually grow or decay depends on which of those particles has a more predominant influence. For those plasmas in thermodynamic equilibrium, the number of faster particles is less than that of slower particles as may be seen from the comparison between (A) and (B) in Fig. 2.7; the plasma wave thus decays. The decay of plasma waves arising from such a resonant coupling between waves and particles is called **Landau damping** (Landau 1946).

As may be clear from the foregoing argument, the decay rate $\gamma_k$ of the amplitude of the plasma wave is proportional to the derivative of the velocity distribution function at the phase velocity. According to the calculation on the basis of RPA (see Appendix 5 for derivation), its rate is given by

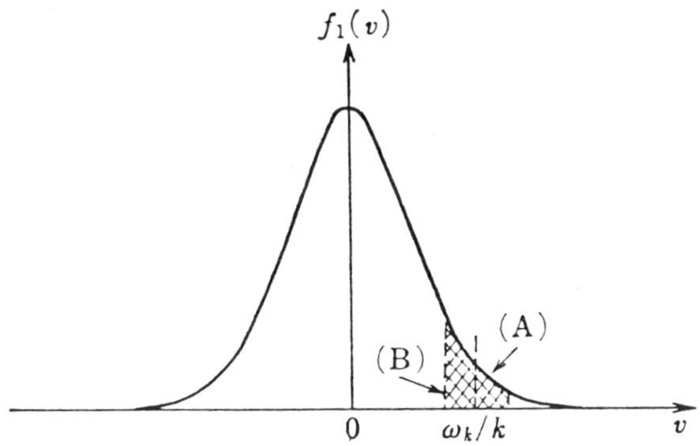


**Figure 2.7.** One-dimensional velocity distribution in the direction of plasma-wave propagation and resonant coupling.

$$\gamma_k = -\frac{\pi\omega_p^2}{2k^2}\,\omega_k\left[\frac{\partial f_1(v)}{\partial v}\right]_{v=\omega_k/k} \tag{2.41}$$

where $f_1(v)$ represents the one-dimensional velocity distribution function in the direction of the wave propagation. It can be calculated from the ordinary three-dimensional velocity distribution function $f(\vec{v})$ according to

$$f_1(v_x) = \int dv_y \int dv_z f(\vec{v}) \tag{2.42}$$

where the x-axis is chosen in the direction of the wave propagation. In the case of the Maxwellian distribution (1.2), we have

$$\gamma_k = \left(\frac{\pi}{8}\right)^{1/2} \omega_p \frac{k_D^3}{k^3} \exp\left(-\frac{M\omega_k^2}{2k^2 k_B T}\right) . \tag{2.43}$$

In the long-wavelength domain satisfying Eq. (2.35), $|\omega_k| \gg \gamma_k$ holds; the amount of Landau damping over a period of the wave is small.

In connection with Eq. (2.41), we note that $\gamma_k$ may become negative if $\partial f_1(v)/\partial v$ takes on a positive value at the phase velocity; the wave amplitude will grow. In thermodynamic equilibrium such does not take place. If a beam of charged particles is injected in the plasma, for example, one can create a distribution in which a sort of "population inversion" is realized and a plasma wave can be amplified. This phenomenon may be looked upon as an **instability.** An instability associated with such a nonequilibrium plasma has been observed and investigated in various plasma phenomena in nature and in the laboratory.

## 2.8 Frequency of Coulomb Collisions

We noted in the previous section a possibility of realizing a plasma-wave instability in a nonequilibrium plasma. Let us then consider the frequency of Coulomb collisions between particles, which would determine the rate at which such a nonequilibrium plasma may approach a state of thermodynamic equilibrium. On the basis of the scattering cross-section (2.3) for momentum transfer, such a frequency $\nu_m(v)$ is given by

$$\nu_m(v) = n\, Q_m v \,. \tag{2.44}$$

For a weakly coupled plasma we may use Eq. (2.15) to find

$$\nu_m(v) = \frac{4\pi n(Ze)^4}{\mu^2 v^3} \ln(9N_D) \,. \tag{2.45}$$

Using a Maxwellian average

$$\langle v^3 \rangle = 8\left(\frac{2}{\pi}\right)^{1/2} \left(\frac{k_B T}{\mu}\right)^{3/2} ,$$

we find the average collision frequency as

$$\nu_m = \left(\frac{\pi}{2}\right)^{3/2} \frac{n(Ze)^4}{\mu^{1/2}(k_B T)^{3/2}} \ln(9N_D) \,. \tag{2.46}$$

It is instructive to compare this result with the plasma frequency, Eq. (2.32). Assuming $\mu = M/2$ for collision between like particles, we find

$$\frac{\nu_m}{\omega_p} = \frac{\sqrt{\pi}}{24N_D} \ln(9N_D) \,. \tag{2.47}$$

As we saw in Eq. (2.12), $N_D$ takes on a huge number for a weakly coupled plasma; hence $\omega_p \gg \nu_m$ holds. In a plasma at high temperature and low density, it is relatively easy to maintain a nonequilibrium state over a time scale on the order of the period of plasma oscillation. A weakly coupled plasma is sometimes referred to as a **collisionless plasma.** The reason may be clear from the comparison of Eq. (2.47); implication is that coupling associated with Coulomb collisions is weak.

On the other hand, for a strongly coupled plasma we obtain from Eq. (2.23)

$$\frac{\nu_m}{\omega_p} \simeq 2\alpha_1{}^2\left(\frac{3}{\pi\Gamma}\right)^{1/2} . \tag{2.48}$$

Even with $\Gamma \geq 1$, $\nu_m \ll \omega_p$ does not always hold true. In comparison with the case of a weakly coupled plasma, the approach to thermodynamic equilibrium through Coulomb collisions is a fast process in a strongly coupled plasma; it is relatively difficult to produce plasma-wave instability.

For an accurate treatment of collision frequencies and transport coefficients, it is necessary to consider a kinetic equation describing a space-time evolution of particle distribution functions. A kinetic-theoretical derivation of the collision term is given in Appendix 3.

## Chapter 3

# Density Correlations in Plasmas

Charged particles in a plasma interact with each other via Coulomb forces, and thereby produce a variety of interesting features as described in the previous chapter. In many cases, these features are expressed in terms of particle-density correlations. For example, the Debye screening and the ion-sphere model are related to the static correlation effects; the plasma oscillation may be looked upon as an appearance of a dynamic correlation effect. The interparticle correlations produced by the Coulomb interaction are thus essentially connected to the plasma properties. In this chapter, we study some of the methods by which those correlation effects may be measured. The density correlations are described in terms of a spectral function called the dynamic structure factor. It is the purpose of the present chapter to clarify the relationship between the dynamic structure factor and a scattering cross-section of a many-particle system.

## 3.1 Propagation of Plasma Oscillation

In Sections 2.5 and 2.7, we considered the propagation

characteristics of the plasma oscillation, a space-charge wave in a plasma. In this section we shall study how such a theoretical prediction is monitored experimentally. Pertinent experiments were carried out by Malmberg and Wharton (1964), Van Hoven (1966), and Derfler and Simonen (1966).

Figure 3.1 is a schematic presentation of the experimental method. A plasma is produced by electric discharge in a container, where grids (A) and (B) are inserted. The distance x between the two grids is variable. By an external signal generator, a sinusoidal potential variation, a $\cos(\omega t)$, is applied to the grid (A); a space-charge wave is thereby excited. The plasma wave, so excited, propagates in the plasma and induces a sinusoidal change* of the potential at the grid (B). Consequently, one can determine the propagation characteristics of the space-charge wave through a measurement of the phase $\phi$ and the amplitude b of the signal received at the grid (B), both as functions of x.

Let us note that this experiment corresponds to a situation slightly different from that analyzed in Section 2.5. We considered in Section 2.5 how the amplitude of a spatially varying charge-density fluctuation with wave number k may vary in time. As a consequence the frequency $\omega_k$ and the decay rate $\gamma_k$ were determined as functions of k in Eqs. (2.38) and (2.41). Mathematically speaking, a complex frequency $\omega_k - i\gamma_k$ was determined for a real wave number k.

In the experiment of Fig. 3.1, on the other hand, a potential variation of a given frequency $\omega$ was applied at the position of the grid (A) and the spatial propagation of its

---

* When a large-amplitude signal is applied, the induced potential deviates from a sinusoidal shape due to nonlinear effects. We shall discuss such a case in Chapter 9.

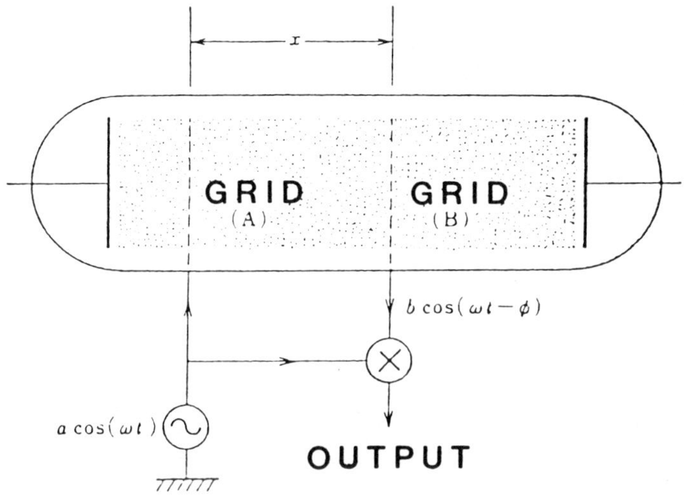


**Figure 3.1.** Experimental setup for space-charge wave propagation (schematic drawing).

amplitude and phase was measured. In these circumstances, the spatial rate of phase change $k_r(\omega) = \partial\phi/\partial x$ and the spatial decay rate $k_i(\omega) = -\partial \ell nb/\partial x$ are determined as functions of $\omega$. Put another way, a complex wave number $k_r(\omega) + ik_i(\omega)$ is determined for a real frequency $\omega$. Although the situations are set differently in those two cases, the physical phenomenon under consideration is the same; one can mutually convert between descriptions $\omega_k - i\gamma_k$ and $k_r(\omega) + ik_i(\omega)$.

The quantities $k_r(\omega)$ and $k_i(\omega)$ may be determined in an experiment of Fig. 3.1 through an interferometric technique by isolating those signals depending only on $\phi$ and b. For example, one obtains an output proportional to $b\cos\phi$ through a time average of the product between a part of the signal applied to the grid (A) and the signal received by the grid (B).

Figure 3.2 shows an example of the output voltage observed in the experiment by Derfler and Simonen at the exciter frequency of 95 MHz. The plasma was produced through ionization

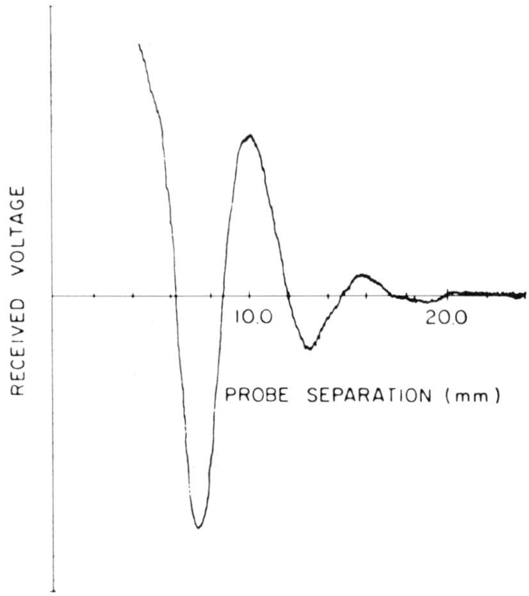


**Figure 3.2.** Interferometer output at 95 Mc/s as a function of grid separation. After Derfler and Simonen (1966).

of sodium vapor; its electron density was $2 \times 10^7$ cm$^{-3}$, temperature 2200 K, and the Debye length 0.7 mm. We computed $N_D \simeq 3 \times 10^4$ and the sodium vapor pressure was $2 \times 10^{-6}$ mmHg; collisional effects between particles may be ignored in the consideration of the propagation characteristics of the plasma oscillation.

The values of $k_r(\omega)$ and $k_i(\omega)$ measured in this way are plotted in Fig. 3.3 by open circles with error bars. The dotted line in the figure shows the thermal-velocity relation, $\omega/k = (3k_BT/m)^{1/2}$, and the chain curve, Bohm-Gross dispersion relation, Eq. (2.39). When $\omega$ is close to $\omega_p$, Eq. (2.39) gives an accurate evaluation, but when $\omega$ is far away from $\omega_p$, one must use a numerical solution to a more general dispersion

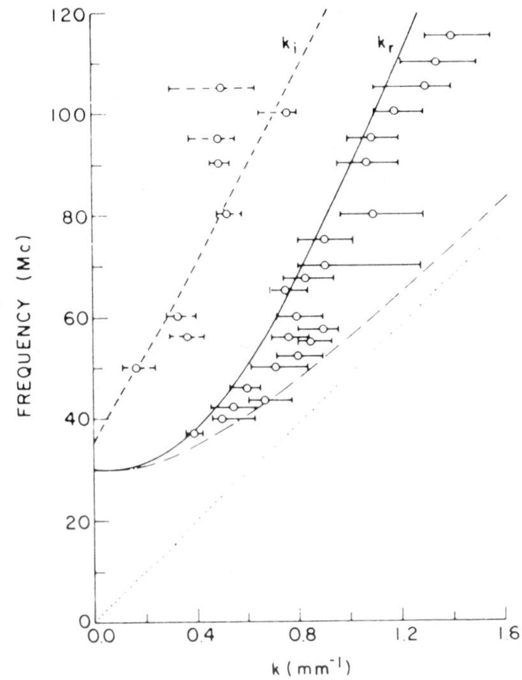


**Figure 3.3.** Dispersion diagram: Circles with bars through them, experimental measurements; $k_r$ and $k_i$, Eqs. (6.15) and (A5.19); and the chain curve, Eq. (2.39). After Derfler and Simonen (1966).

relation for $k_r(\omega)$ and $k_i(\omega)$. The real and dashed curves in Fig. 3.3 represent results of such a numerical solution on the basis of the RPA dielectric response function obtained in Appendix 5. We see that the experimental results agree well with the theoretical prediction.

## 3.2 Cross-Section of Thomson Scattering

In the experiment mentioned in the previous section, the spatial correlation of the potential field associated with propagation of a space-charge wave was measured through signals received by the grids. In many of the plasma experiments, however, one finds cases in which it is practically infeasible to introduce an external probe such as a grid. Even if a probe can be introduced, the state of the plasma under investigation may be awfully disturbed by so doing; it is likely that the inherent errors would be substantial.

The scattering of an electromagnetic wave, which is to an extent devoid of such difficulties, can directly reveal features of the density correlations in the plasma. To elucidate the principle, we begin with a consideration of the cross-section of individual charged particles against scattering of the electromagnetic wave. In this chapter, we assume that the relativistic and quantum effects are negligible as far as the motion of charged particles is concerned. Some of the relativistic effects will be treated in Chapter 11.

We consider a particle with electric charge Ze and mass M located in the plane-wave electromagnetic field, whose electric field is expressed as

$$\vec{E}(\vec{r}, t) = \vec{E}_1 \cos(\vec{k}_1 \cdot \vec{r} - \omega_1 t) . \tag{3.1}$$

The quantities $\omega_1$ and $k_1$ satisfy $\omega_1/k_1 = c$, where $c = 2.9978 \times 10^{10}$ cm/s is the light velocity in vacuum. The charged particle obeys the equation of motion,

$$\ddot{\vec{r}} = \frac{Ze}{M} \vec{E} .$$

The temporal variation of the dipole moment $\vec{d}$ is given by

$$\ddot{\vec{d}} = Ze\ddot{\vec{r}} = \frac{(Ze)^2}{M} \vec{E} . \tag{3.2}$$

Since the rate of energy emitted by an electric dipole is $2(\ddot{d})^2/3c^3$, the emission rate from the charged particle subjected to the electromagnetic field (3.1) is calculated as

$$\frac{2}{3c^3}\frac{(Ze)^4}{M^2}E^2 = \frac{8\pi}{3}\left[\frac{(Ze)^2}{Mc^2}\right]^2\left(\frac{E^2}{4\pi}\right)c \ . \qquad (3.3)$$

The incident plane wave has its total energy density of $E^2/4\pi$; the energy density associated with the electric field is $E^2/8\pi$ and the magnetic field has the same amount of energy density. As $(E^2/4\pi)c$ is the energy flux of the incident plane wave, we find the scattering cross-section of a charged particle from Eq. (3.3) as

$$\sigma = \frac{8\pi}{3}\frac{(Ze)^4}{M^2c^4} \ . \qquad (3.4)$$

The quantity $e^2/mc^2 = 2.818 \times 10^{-13}$ cm is the classical radius of an electron; corresponding to Eq. (3.4),

$$\sigma_T = \frac{8\pi}{3}\left(\frac{e^2}{mc^2}\right)^2 = 6.653 \times 10^{-25} \text{ cm}^2 \qquad (3.5)$$

is called the **cross-section of Thomson scattering.** As Eq. (3.4) indicates, the electromagnetic wave cross-section of a charged particle is inversely proportional to the square of the mass. In an ordinary gaseous plasma, the mass of an ion is at least 1837 times as large as that of an electron; as far as the scattering of electromagnetic waves is concerned, the cross-section of electrons is predominant. If the electro-magnetic waves scattered by individual electrons do not interfere with each other, that is, if the correlation effects between electrons are negligible, the total cross-section of the plasma against scattering of the electromagnetic wave is proportional to the number of electrons in the scattering volume. In these circumstances one can deduce the electron density from a measurement of the total cross-section of scattering.

### 3.3 Doppler Effect

The total cross-section of an electron against scattering of the electromagnetic wave is given by the cross-section of Thomson scattering, Eq. (3.5). The frequency of the scattered wave is generally different from that of the incident wave, however; one of the reasons is the **Doppler effect** associated with the motion of electrons.

Let $\vec{v}$ be the velocity of an electron in the laboratory system (see Fig. 3.4). As we find by substituting the orbital $\vec{r} = \vec{r}(t = 0) + \vec{v}t$ of the electron motion in Eq. (3.1), the electron perceives $\omega' = \omega_1 - \vec{k}_1 \cdot \vec{v}$ as the frequency of the incident wave. Such a motional change of the frequency is called the Doppler effect. The electron moving with the velocity $\vec{v}$ emits a dipolar electromagnetic wave with frequency $\omega'$ in

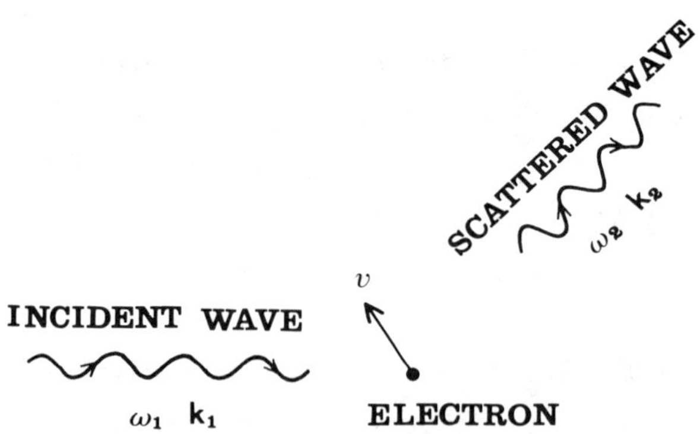


**Figure 3.4.** Scattering of electromagnetic wave by an electron with velocity $\vec{v}$.

its own frame of reference. We then observe its component radiated in the direction of the wave vector $\vec{k}_2$ by a detector fixed to the laboratory system. The frequency of the received electromagnetic wave satisfies $\omega_2 = ck_2$. Using the relation of the Doppler effect once again, we find $\omega_2 = \omega' + \vec{k}_2 \cdot \vec{v}$. Consequently, the frequency shift between the incident and the scattered waves is given by

$$\omega_2 - \omega_1 = \vec{v} \cdot (\vec{k}_2 - \vec{k}_1) \ . \tag{3.6}$$

If the incident wave is monochromatic with the frequency $\omega_1$, the frequency $\omega_2$ of the scattered wave has a characteristic spread corresponding to the velocity distribution of the electrons. For a Maxwellian distribution (1.2) with temperature T, the frequency spectrum of the scattered wave takes the form as shown in Fig. 2.5 where

$$\omega = \omega_2 - \omega_1 \ , \tag{3.7}$$

$$\vec{k} = \vec{k}_2 - \vec{k}_1 \ . \tag{3.8}$$

In these circumstances, the electron temperature can be determined from the spread in the frequency spectrum of the scattered waves. Equations (3.7) and (3.8) are important relations which generally appear in the scattering cross-section of a many-particle system to be discussed later; they correspond to the conservation laws of energy and momentum in the event of scattering.

## 3.4 Density Correlations and Scattering Cross-Sections

Toward the end of Section 3.2 we spoke about a proviso that correlations between electrons be negligible. As the title of the present chapter suggests, however, the density correlations exert important influences on many of the plasma phenomena. In this section we consider the effect of density correlations between electrons on the cross-section against scattering of the electromagnetic wave.

We assume, as shown in Fig. 3.5, that a plane wave with frequency $\omega_1$ and wave vector $\vec{k}_1$ is incident to the plasma, and a scattered wave with a frequency $\omega_2$ is detected in the direction of $\vec{k}_2$. Naturally, $\omega_1/k_1 = \omega_2/k_2 = c$ hold. The intensity of the scattered wave from an infinitesimal volume element $d\vec{r}_1$ around $\vec{r}_1$ is proportional to $\rho(\vec{r}_1, t_1)d\vec{r}_1$, where $\rho(\vec{r}_1, t_1)$ represents the electron density in the plasma at $\vec{r}_1$ and $t_1$. The advance in the phase of the incident wave up to $\vec{r}_1$ and $t_1$ is (apart from an immaterial constant term) $\vec{k}_1 \cdot \vec{r}_1 - \omega_1 t_1$; the phase advance of the wave between the scattering point $\vec{r}_1$ and $t_1$ and the receiving point $\vec{R}$ and t is $\vec{k}_2 \cdot (\vec{R} - \vec{r}_1) - \omega_2(t - t_1)$. Hence, neglecting a constant term $\vec{k}_2 \cdot \vec{R}$ again, we write

$$\text{(scattered wave)} \sim \int_V d\vec{r}_1 \rho(\vec{r}_1, t_1) \exp[i(\vec{k}_1 - \vec{k}_2) \cdot \vec{r}_1 - i(\omega_1 - \omega_2)t_1 - i\omega_2 t] . \tag{3.9}$$

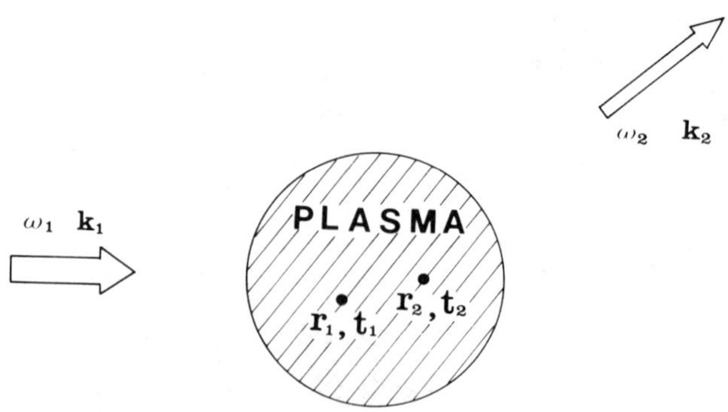


**Figure 3.5.** Scattering of electromagnetic wave by a plasma.

Since $\rho(\vec{r}_1, t_1)$ is a fluctuating stochastic variable, this formula implies an amplitude modulation by an amount of

$$\int_V d\vec{r}_1 \rho(\vec{r}_1, t_1) \exp[i(\vec{k}_1 - \vec{k}_2) \cdot \vec{r}_1 - i(\omega_1 - \omega_2)t_1] \tag{3.10}$$

to an electromagnetic wave with frequency $\omega_2$.

To decompose the stochastically fluctuating amplitude into its spectral components, we calculate the mean square average of Eq. (3.10):

$$\langle|\text{amplitude}|^2\rangle \sim \int_V d\vec{r}_1 \int_V d\vec{r}_2 \langle \rho(\vec{r}_1, t_1)\rho(\vec{r}_2, t_2)\rangle$$

$$\cdot \exp[i(\vec{k}_1 - \vec{k}_2) \cdot (\vec{r}_1 - \vec{r}_2) - i(\omega_1 - \omega_2)(t_1 - t_2)] . \tag{3.11}$$

Assuming that the plasma in the scattering volume V is uniform and stationary, we may take the space and time correlation function $\langle \rho(\vec{r}_1, t_1)\rho(\vec{r}_2, t_2)\rangle$ of the density fluctuations as a function of only the variables, $\vec{r}_1 - \vec{r}_2$ and $t_1 - t_2$. In terms of the Fourier components, we express

$$\langle \rho(\vec{r}_1, t_1)\rho(\vec{r}_2, t_2)\rangle = \frac{1}{V} \sum_{\vec{k}} \int d\omega S(\vec{k}, \omega) \exp[i\vec{k} \cdot (\vec{r}_1 - \vec{r}_2)$$

$$- i\omega(t_1 - t_2)] . \tag{3.12}$$

The spectral function $S(\vec{k}, \omega)$ of the electron-density fluctuations introduced here is called the **dynamic structure factor** or the **dynamic form factor**; it is an important quantity in the treatment of the plasma many-body problem. In the following section we summarize essential relations with regard to this function.

Substituting Eq. (3.12) in Eq. (3.11) and integrating the result with respect to $\vec{r}_1$ and $\vec{r}_2$, we obtain

$$<|\text{amplitude}|^2> \sim V \int d\omega S(\vec{k}, \omega) \tag{3.13}$$

with the aid of the conditions (3.7) and (3.8). Hence $S(\vec{k}, \omega)$ represents the frequency spectrum of the amplitude-modulated scattered waves.

Consideration of Eq. (3.9) combined with Eq. (3.13) may convince us that the differential cross-section of the plasma against scattering of the electromagnetic wave is proportional to $S(\vec{k}, \omega)$. To determine the constant of proportionality, we must carry out the detailed calculations, which are described in Appendix 9. We quote the result in the following.

Denoting by $\hat{n}_1$ and $\hat{n}_2$ the unit vectors in the directions of the electric-field polarization for the incident and scattered waves, we express the differential cross-section of scattering into a solid angle $d\Omega$ in the $\vec{k}_2$ direction and into a frequency interval between $\omega_2$ and $\omega_2 + d\omega$ as

$$\frac{d^2Q}{d\Omega d\omega} = V\left(\frac{e^2}{mc^2}\right)^2 (\hat{n}_1 \cdot \hat{n}_2)^2 S(\vec{k}, \omega) . \tag{3.14}$$

The incident and scattered waves are connected via Eqs. (3.7) and (3.8) with $\omega$ and $\vec{k}$.

When no attention is paid to the polarization directions in the experiment, Eq. (3.14) is averaged over the two states of $\hat{n}_1$ and is summed over $\hat{n}_2$, with the result,

$$\frac{d^2Q}{d\Omega d\omega} = V\left(\frac{e^2}{mc^2}\right)^2 \left(1 - \frac{1}{2}\sin^2\theta\right) S(\vec{k}, \omega) , \tag{3.15}$$

where $\theta$ is the scattering angle between $\vec{k}_1$ and $\vec{k}_2$.

As may be clear from those formulas, the spectral distribution of the scattered wave takes a form identical to the frequency spectrum of electron-density fluctuations. In the short-wavelength domain such that $k = |\vec{k}_2 - \vec{k}_1|$ satisfies $k^2 \gg k_D^{\,2}$, the individual-particlelike fluctuations are the main constituent of $S(\vec{k}, \omega)$; the spectrum takes the form of Fig. 2.5. In this case the superposition of Doppler effects arising from the individual electrons, as mentioned in Section 3.3, is applicable.

When $k^2 \ll k_D^{\,2}$, on the other hand, the fluctuations behave collectively; the spectrum now takes the form of Fig. 2.4. One anticipates in these circumstances that the scattered wave may be split into two sideband waves, $\omega_2 = \omega_1 \pm \omega_p$. Those differences carry important consequences in the interpretation of the scattering data from the ionosphere, to be treated in Chapter 4.

## 3.5 Structure Factors and the Radial Distribution Function

The dynamic structure factor introduced in Eq. (3.12) represents the Fourier transform of the space and time correlation function of the density fluctuations,*

---

* In this section we assume the plasma to be translationally invariant both in space and in time.

$$S(\vec{k}, \omega) = \frac{1}{2\pi} \int_{-\infty}^{\infty} d(t_1 - t_2) \int_{\text{unit volume}} d(\vec{r}_1 - \vec{r}_2)$$

$$\langle \rho(\vec{r}_1, t_1)\rho(\vec{r}_2, t_2) \rangle$$

$$\cdot \exp[-i\vec{k} \cdot (\vec{r}_1 - \vec{r}_2) + i\omega(t_1 - t_2)] \ . \tag{3.16}$$

Substituting Eq. (2.24) in Eq. (3.16) and rewriting the result in terms of the spatial Fourier component (2.26) of the fluctuations, we obtain

$$S(\vec{k}, \omega) = \frac{1}{2\pi} \int_{-\infty}^{\infty} dt \langle \rho_{\vec{k}}(t' + t)\rho_{-\vec{k}}(t') \rangle \exp(i\omega t) \ . \tag{3.17}$$

This expression takes a form of the **Wiener-Khinchine theorem** familiar in the noise theory. We may further derive from Eq. (3.16)

$$\sum_{\vec{k}} \int_{-\infty}^{\infty} d\omega S(\vec{k}, \omega) = \langle \{\rho(\vec{r}, t)\}^2 \rangle \ , \tag{3.18}$$

corresponding to the **Parseval theorem.** We observe in those formulas that $S(\vec{k}, \omega)$ may be interpreted as the spectral function of density fluctuations decomposed in both the wave number and frequency domains.

Integration of the dynamic structure factor over the entire frequency domain yields the **static structure factor** $S(\vec{k})$ corresponding to the spectrum of density fluctuations in the wave number space. Its definition then is

$$S(\vec{k}) \equiv \frac{1}{n} \int_{-\infty}^{\infty} d\omega \; S(\vec{k}, \omega) \;, \tag{3.19}$$

and can be expressed from Eq. (3.17) as

$$S(\vec{k}) = \frac{1}{n} \langle |\rho_{\vec{k}}(t)|^2 \rangle \; . \tag{3.20}$$

The static structure factor appears frequently in the theory of liquid and in the analysis of X-ray scattering data. In fact, if we are interested in the scattering of the electromagnetic wave into a solid angle $d\Omega$ only, without delving into decomposition of frequencies, such a differential cross-section is given by

$$\frac{dQ}{d\Omega} = N\left(\frac{e^2}{mc^2}\right)^2 \left(1 - \frac{1}{2} \sin^2\theta\right) S(\vec{k}) \;, \tag{3.21}$$

an integration of Eq. (3.15) over the frequency $\omega$. Here $N = nV$ refers to the total number of electrons in the scattering volume.

An important physical quantity directly related to $S(\vec{k})$ is the radial distribution function $g(\vec{r})$. The meaning of this function is as follows: $g(\vec{r})$ is the probability of finding another particle in the characteristic volume $1/n$ around $\vec{r}$ provided that a particle is already found at the origin ($\vec{r} = 0$). It follows that if no correlations exist between particles, then $g(\vec{r}) = 1$. $S(\vec{k})$ and $g(\vec{r})$ are related to each other via the Fourier transformation,

$$g(\vec{r}) - 1 = \frac{1}{n} \sum_{\vec{k}} [S(\vec{k}) - 1] \exp(i\vec{k} \cdot \vec{r}) \; . \tag{3.22}$$

Consequently if no correlations exist between particles, then $S(\vec{k}) = 1$ as well.

**Chapter 4**

# Radio-Wave Scattering from the Ionosphere

Around the year 1960 a series of experiments were performed to probe the state of the ionospheric plasma by means of back-scattering technique of radar pulses. It was originally conceived that the time-dependence of the scattered signal would reveal directly the electron density and temperature profiles of the ionosphere as functions of distance from the surface of the earth. Unexpectedly, however, the measured results taught us the importance of correlation effects between the charged particles in the ionosphere. It was particularly a remarkable observation that the spectrum of the scattered waves carried a strong influence of the ionic motion despite the fact that the electrons were the scatterers. Taking those as experimental examples, let us specifically consider how the correlations between charged particles influence the cross-section of scattering.

## 4.1 Propagation of Electromagnetic Waves

Before taking up the scattering problem of electromagnetic wave in the ionosphere, it is essential first of all to understand how the electromagnetic plane-wave

$$\vec{E}(\vec{r}, t) = \vec{E}_0 \cos(\vec{k} \cdot \vec{r} - \omega t)$$

$$\vec{B}(\vec{r}, t) = \vec{B}_0 \cos(\vec{k} \cdot \vec{r} - \omega t) \tag{4.1}$$

propagates in the plasma. The Maxwell equations read

$$\nabla \times \vec{E} + \frac{1}{c}\frac{\partial \vec{B}}{\partial t} = 0 , \tag{4.2}$$

$$\nabla \times \vec{B} - \frac{1}{c}\frac{\partial \vec{E}}{\partial t} = \frac{4\pi}{c}\vec{j} , \tag{4.3}$$

where $\vec{j}$ is the electric current density. Substitution of Eqs. (4.1) in Eq. (4.2) yields

$$\vec{k} \times \vec{E}_0 = \frac{\omega}{c}\vec{B}_0 . \tag{4.4}$$

The electric current density is calculated as

$$\vec{j} = n\,Ze\dot{\vec{r}} = -\frac{(Ze)^2 n}{M\omega}\vec{E}_0 \sin(\vec{k} \cdot \vec{r} - \omega t) \tag{4.5}$$

from the equation of motion, $M\ddot{\vec{r}} = Ze\vec{E}$, of a charged particle in the plasma. Finally substituting Eqs. (4.1) and (4.5) in Eq. (4.3), we obtain

$$\vec{k} \times \vec{B}_0 + \frac{\omega}{c}\vec{E}_0 = \frac{\omega_p^2}{c\omega}\vec{E}_0 \tag{4.6}$$

where $\omega_p$ is the plasma frequency defined in Eq. (2.32).

We next eliminate $\vec{B}_0$ by substituting Eq. (4.4) in Eq. (4.6) to find

$$c^2\vec{k} \times (\vec{k} \times \vec{E}_0) + (\omega^2 - \omega_p^2)\vec{E}_0 = 0 . \tag{4.7}$$

We may decompose Eq. (4.7) by choosing $\vec{E}_0$ in the x direction and $\vec{k}$ in the z direction, for example; the condition that Eq. (4.7) has a solution with $\vec{E}_0 \neq 0$ is then found to be

$$\omega^2 = \omega_p{}^2 + c^2k^2 . \tag{4.8}$$

This formula gives a dispersion relation between the frequency $\omega$ and the wave number k of the electromagnetic wave in the plasma. Equation (4.8) represents a dispersion relation for the transverse (electromagnetic) wave, while Eqs. (2.38) and (2.39) are the dispersion relations for the longitudinal (space-charge) wave.

The refractive index of a dielectric medium such as a plasma is the ratio of the light velocity in a vacuum to the phase velocity of an electromagnetic wave in the medium; the square of the refractive index then gives the dielectric constant $\varepsilon_T(\omega)$ for propagation of the electromagnetic wave. Consequently, rewriting Eq. (4.8) as

$$\frac{\omega^2}{k^2} = \frac{c^2}{\varepsilon_T(\omega)} , \tag{4.9}$$

we find

$$\varepsilon_T(\omega) = 1 - \frac{\omega_p{}^2}{\omega^2} . \tag{4.10}$$

For $\omega > \omega_p$, $1 > \varepsilon_T(\omega) > 0$, so that the electromagnetic wave in the plasma propagates at a phase velocity faster than the light velocity in a vacuum. For $\omega < \omega_p$, on the other hand, Eq. (4.10) indicates that $\varepsilon_T(\omega) < 0$; for a real $\omega$, Eq. (4.9) gives an imaginary value of k. This implies that an electromagnetic wave with $\omega < \omega_p$ cannot propagate in the plasma. The following example elucidates the situation.

Let us consider the case of Fig. 4.1 where the plasma density n(x) increases gradually in the x direction and an electromagnetic wave with frequency $\omega$ is injected from the left.* The plasma frequency, $\omega_p(x) = [4\pi(Ze)^2 n(x)/M]^{1/2}$, is now a function of x. As this value increases with x, the wavelength $2\pi/k$ of the electromagnetic wave correspondingly increases in accord with Eqs. (4.9) and (4.10); it would diverge at the (cutoff) point satisfying $\omega_p(x) = \omega$. The electromagnetic field in this vicinity exhibits almost no spatial variations and varies sinusoidally in time at the frequency $\omega$. Since k becomes imaginary toward the right of the cutoff point, the wave cannot propagate into this domain; the electromagnetic wave is reflected toward the left at this point.

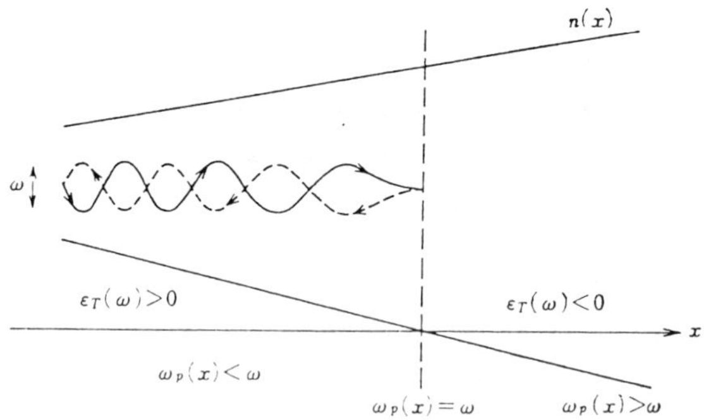


**Figure 4.1.** Cutoff of an electromagnetic wave.

---

* It is here assumed that the relative change of the density over the wavelength, $|dn/dx|(2\pi c/\omega n)$, is much smaller than unity.

As the foregoing consideration indicates, the plasma frequency gives the **cutoff frequency** for the propagation of an electromagnetic wave in the plasma; an electromagnetic wave with a frequency lower than this cannot propagate through the plasma. In the case of the ionosphere, it is also called the **critical frequency.** Radio communication uses this feature in that the electromagnetic waves with frequencies less than the critical frequency of the ionosphere are reflected back to the earth.

In the analysis described above, we have implicitly assumed that the effects of an external magnetic field and the thermal motion of plasma particles are negligible. When a uniform magnetic field B is applied externally to the plasma, as we shall stn Chapter 5, a charged particle performs a periodic motion around the magnetic lines of force at the **cyclotron frequency**

$$\omega_H = \frac{|Z|eB}{Mc} . \tag{4.11}$$

If the magnetic field is sufficiently weak so that $\omega >> \omega_H$ and $\omega_p >> \omega_H$ hold, the effect of the magnetic field can be ignored for propagation of the electromagnetic wave in the plasma. (Propagation of electromagnetic waves in a magnetized plasma will be treated later in Chapter 8.) For those nonrelativistic plasmas in which the thermal motion is considerably smaller than the light velocity, one can likewise ignore the effect of thermal motion on the propagation of electromagnetic waves.

As we have seen in Fig. 4.1, the propagation of an electromagnetic wave is cut off when $\varepsilon_T(\omega)$ changes its sign from positive to negative through

$$\varepsilon_T(\omega) = 0 \ . \tag{4.12}$$

Another possibility exists, however, for a change of the sign in $\varepsilon_T(\omega)$: It may diverge to $+\infty$ in a positive domain of $\varepsilon_T(\omega)$ and then enter a negative domain through $-\infty$. Such a case is called a **resonance**, where a phenomenon entirely different from the cutoff is expected. The frequency satisfying

$$\varepsilon_T(\omega) = \infty \tag{4.13}$$

is then called the **resonance frequency.**

In the case of the dielectric constant as expressed in Eq. (4.10), such a resonance would not take place. When an external magnetic field is applied to a plasma, for example, Eq. (4.13) can be satisfied at the cyclotron frequency (4.11). This is the **cyclotron resonance**, to be considered in Chapter 8.

To illustrate the difference between the resonance condition (4.13) and the cut-off condition (4.12), we take up another example illustrated in Fig. 4.2. The wavelength $2\pi/k$ of an electromagnetic wave with frequency $\omega$ injected from the left will gradually decrease in accord with Eq. (4.9). The phase velocity $\omega/k$ of the electromagnetic wave likewise approaches zero; it takes a long time for the wave to reach the resonance point $x_R$. Generally in a physical medium such as a plasma, there always exists a mechanism such as the collisional effect which acts to absorb the wave energy and turns it into heat. Such a dissipative effect, weak as it may be, would bring about a substantial absorption effect if an electromagnetic wave stays there for a long time. In addition, since the phase velocity of the wave diminishes, the resonant interaction between wave and particles as considered in Section 2.7 begins

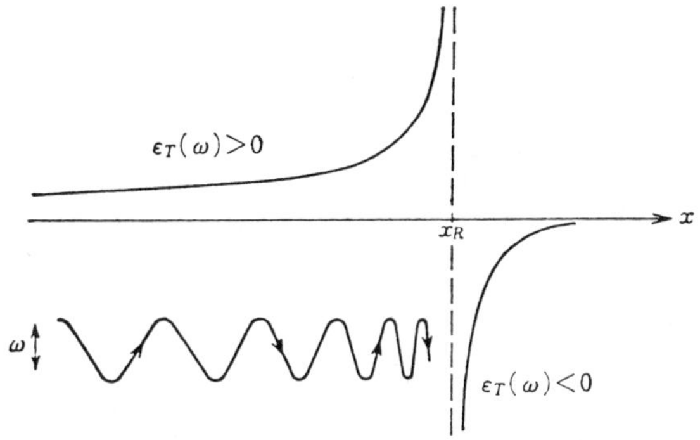


**Figure 4.2.** Resonance of an electromagnetic wave.

to produce a sizable absorption effect. Consequently an electromagnetic wave is strongly absorbed in the vicinity of the resonance point. In the case of the cut-off condition (4.12), on the other hand, a wave is reflected almost without absorption; hence the phenomenological difference is substantial.

## 4.2 Experiments by Bowles

The F layer of the ionosphere is located at an altitude of 200 - 500 km; its electron density and temperature are estimated to be $10^5$ - $10^6$ cm$^{-3}$ and approximately 1500 K. Assuming the maximum value of the electron density to be around $2 \times 10^6$ cm$^{-3}$, we find the corresponding plasma frequency to be 12.7 MHz. This is the critical frequency; an electromagnetic wave with a frequency less than that will be reflected by the F layer. An electromagnetic wave with a frequency exceeding the critical frequency, on the other hand, passes through the F layer; its tiny fraction, however, is scattered by the electron-density fluctuations and is sent back toward the surface of the earth.

In 1958, Bowles (1958, 1961) sent radar pulses vertically

upwards and measured weak noiselike signals scattered back from the F layer. The transmission frequency of the radar was 40.92 MHz, the width of the frequency spread 9 kHz, the pulse width 120 μs, the pulse repetition rate 25 - 40 $s^{-1}$, and the peak power of the pulses 1 MW for most of the experiments. The frequency bandwidth of the receiver detecting the scattered signal was 9, 13, or 35 kHz; its central frequency was the same as the transmitter frequency or could be adjusted around it. For an electromagnetic wave scattered at the altitude of 300 km, the round-trip time is 2 ms; the receiver signal of the scattered wave can be regarded as isolated from the transmission pulses.

The electron density of the ionosphere varies with the altitude. If the strength of a scattered wave is regarded as proportional to the electron density, following the argument of Section 3.2, the temporal variation on the strength of the receiver signal should reveal the electron-density profile. Figure 4.3 shows a result of such an observation. The altitude at which the electron density takes on its maximum value is approximately 330 km; the profile generally agrees with the data obtained by other methods.

The absolute strengths of scattered waves observed in these experiments were in good agreement with those expected from the cross-section of Thomson scattering discussed in Section 3.2. This can be examined through comparison between the maximum value of the electron density determined by the critical frequency and the maximum strength of the scattered wave. When the critical frequency was 13 MHz the maximum strength of the receiver signal was $4.3 \times 10^{-16}$ W; this was almost equal to the value calculated on the basis of the Thomson scattering cross-section.

It thus appeared that the correspondence between the

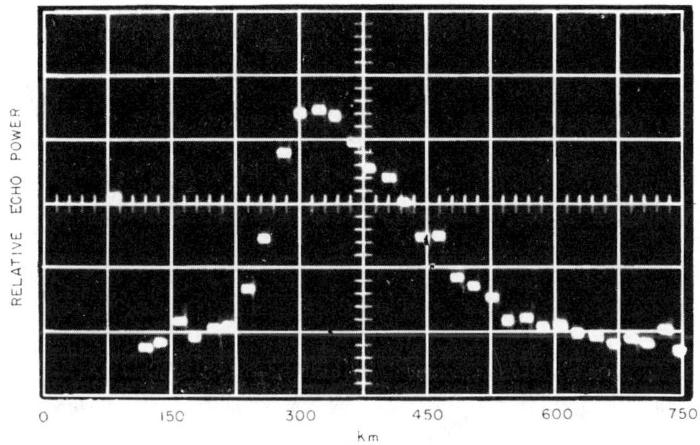


**Figure 4.3.** Radar-backscattering measurement of the ionospheric electron distribution as a function of the scattering altitude. [1959 Feb. 27, 7:40 p.m.--Illinois local time.] After Bowles (1961).

strength of the receiver signal and the electron density in the ionosphere was well explained in terms of Thomson scattering by individual electrons (Gordon 1958). A difficulty arose when one attempted to interpret the spectral character of the scattered waves. Let us first look at the observational results: There was no change observed in the strength of receiver signal when the frequency bandwidth was increased from 9 kHz to 35 kHz. Next, when the central frequency of the receiver was shifted by 15 kHz above or below the transmitter frequency with the frequency bandwidth fixed at 9 kHz, no signals of the scattered waves were detected. Those results implied that the spectral width of the scattered waves exhibited almost no increase above 9 kHz, which was the original frequency spread in the transmitter signal.

Recalling the analysis of Section 3.3, let us compute the magnitude of the frequency spread expected from the Doppler effect due to thermal motion of electrons. In the case of back scattering from the ionosphere the wave vectors of the incident and scattered waves have the relation, $\vec{k}_1 = -\vec{k}_2$; we thus compute $k = 2k_1 = 1.7 \times 10^{-2}\ \text{cm}^{-1}$ with the aid of Eq. (3.8). Consequently the frequency spread $(k/\pi)(k_B T/m)^{1/2}$ due to the thermal motion takes on a value 82 kHz at T = 1500 K. As the comparison in Fig. 4.4 clearly indicates, those computed values are obviously in contradiction with the experimental results stated above. It was raised as a big question to explain why the spectral spread expected from the thermal motion of electrons did not show up in the observation results.

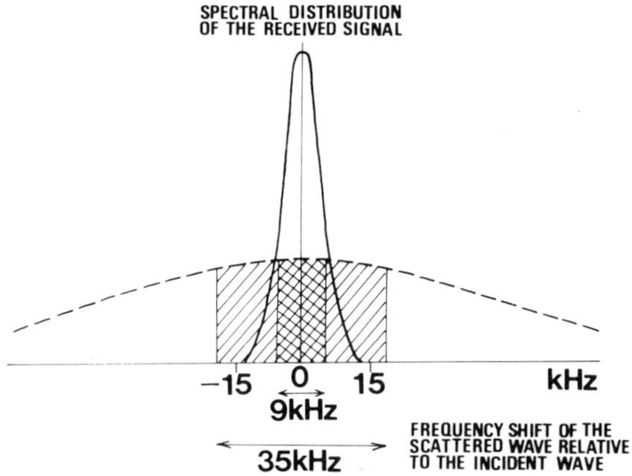


**Figure 4.4.** Interpretation of the spectral distribution of the received signal in Bowles' back-scattering experiment. The solid and dashed curves represent spectral distributions with widths 9 kHz and 82 kHz, respectively. If the received signal had a spectral distribution of the dashed curve, there would have been a substantial difference between the intensities of the received signal when the receiver bandwidth was 9 kHz (the doubly hatched area) and when the receiver bandwidth was 35 kHz (the hatched area); a shift of the receiver central frequency by ± 15 kHz would not have resulted in a significant change in the intensities of the received signal. All of these were contrary to the observed results.

## 4.3 Collective Versus Individual-Particle Aspects of Scattering

The Debye wave number of electrons plays an essential part in the theoretical interpretation of the results of the scattering experiments. Substituting $Z = -1$, $n = 2 \times 10^6$ cm$^{-3}$, and $T = 1500$ K in Eq. (2.38), we find $k_D = 5.3$ cm$^{-1}$. Comparing this with the evaluation of k in the previous section, we clearly see that the condition (2.35), $k^2 \ll k_D^2$, is satisfied. Put another way, the scattering experiment by Bowles detected what amounted to be the collective behavior of electron-density fluctuations in the ionosphere, which is distinctively different from the individual-particlelike behavior. Theoretical analysis predicts the spectral shape in the form of Fig. 2.4, not in the form of Fig. 2.5.

At this stage we encounter another difficulty. This is related to the fact that the total spectral strength contained in Fig. 2.4 is only a small fraction $k^2/k_D^2$ ($\simeq 10^{-5}$) of that contained in Fig. 2.5. For, in the long-wavelength domain where Fig. 2.4 is applicable, the effect of Debye screening (Section 2.2) is so predominant that the strength of density fluctuations is almost completely suppressed. In fact, a theoretical calculation for a weakly coupled, classical one-component plasma in thermodynamic equilibrium (see Appendix 7) shows that the static structure factor Eq. (3.19) is given by

$$S(k) = \frac{k^2}{k^2 + k_D^2} \, . \tag{4.14}$$

In the short-wavelength domain $k^2 \gg k_D^2$, one finds $S(k) \simeq 1$, indicating a state with uncorrelated superposition of individual particles. In the long-wavelength domain $k^2 \ll k_D^2$, on the contrary, the correlation effects are so large that S(k) $\simeq k^2/k_D^2$. Since S(k) is the integral of S(k, ω) with respect to ω, it is proportional to the total strength of the scattered electromagnetic wave into a given direction, as shown in Eq. (3.21). Numerical examination in the previous section indicated that, if we adopted an individual-particlelike scattering model of Fig. 2.5, we could explain the observational data at least for the total strength of the scattered waves. With the collective scattering model of Fig. 2.4 at hand, however, this explanation had to be given up, because the total theoretical cross-section was too small to account for the observation.

Let us note in this connection that the plasma frequency corresponding to the electron density $n = 2 \times 10^6$ cm$^{-3}$ is approximately 13 MHz, which is by far a high frequency entirely out of the scale in Fig. 4.4. As Fig. 2.4 shows, the spectral strength of the collective fluctuations is concentrated in the vicinity of $\omega = \pm \omega_p$; no strength remains in the vicinity of $\omega = 0$. Since $k^2 \ll k_D^2$, we cannot escape from consideration of the collective aspect in the scattering. Then it becomes impossible to explain either the total strength or the frequency spectrum of the observed scattered waves on the basis of such a collective model. An essential point must have been overlooked in the theoretical procedure of interpretation mentioned above.

## 4.4 Dressed Particles

In ordinary plasmas, not only the electrons but also the ions constitute the mobile charged particles; an OCP model is not strictly applicable. As we have examined in Section 3.2, the cross-section of an electron is overwhelmingly large as compared with that of an ion as far as scattering of the electromagnetic wave is concerned. In the treatment of scattering phenomena described in Section 4.2, we might legitimately regard the ionosphere as an OCP of electrons in which the ions acted only as a neutralizing space-charge background; the argument in the previous section was based on such an assumption.

The ions in real plasmas are undoubtedly discrete particles. Hence, those ions do interact with other charged particles via Coulomb forces and thereby produce correlation effects such as screening. The screening effect is particularly dominant for $k^2 \ll k_D{}^2$, and is thus essential in the interpretation of the scattering experiment of Section 4.2.

Let us pick an ion in the plasma and consider those electrons which contribute to the screening of the ionic Coulomb field. The scattering cross-section for each of those electrons while following the motion of the ion takes on the value as given by Eq. (3.5). Since the velocity of the ionic thermal motion is much smaller than that of electrons, one expects that the spectral spread due to the Doppler effect may likewise assume a much reduced value. In the case of Bowles' experiment, the thermal spread due to hydrogen ions is only 1.9 kHz; for oxygen ions, a dominant constituent of the ionosphere, this value is further reduced. An ionic Doppler spread by an amount of 1.9 kHz at most would thus be added to the solid curve of Fig. 4.4 representing the original frequency spread of the transmitted radar wave; this would not produce a significant change in its shape. The observed results were indeed in agreement with such a spectral shape.

In the treatment of many-particle systems such as plasmas, one frequently introduces the idea of a **dressed particle** consisting of a given particle plus its correlated particle cloud, and thereby takes an effective account of the salient correlational effects (Thompson and Hubbard 1960; Hubbard 1961; Pines and Nozières 1966; Ichimaru 1973). The foregoing interpretation of the scattering cross-section is an application of such a notion of dressed particles. (See Appendix 8 for mathematical discussion on the dressed-particle concept.)

To appreciate the features of screening in a two-component plasma consisting of electrons and ions, let us revisit the problem considered in Section 2.2. We use the subscripts, e and i, to distiguish between the quantities associated with electrons and ions. The condition for space-charge neutralization on average then specifies $n_i = n_e/Z$ where $Ze$ is the ionic charge. Extending the analysis of Section 2.2 to the case of a two-component plasma, we find that the Debye length $\lambda_D$ in the effective potential Eq. (2.9) is given by

$$\frac{1}{\lambda_D^2} = k_{De}^2 + k_{Di}^2 , \tag{4.15}$$

where

$$k_{De}^2 = \frac{4\pi n_e e^2}{k_B T_e} , \quad k_{Di}^2 = \frac{4\pi n_i (Ze)^2}{k_B T_i} , \tag{4.16}$$

$T_e$ and $T_i$ represent the temperatures of the electron and ion systems, respectively. The quantities $k_{De}$ and $k_{Di}$ of Eq. (4.16) are the Debye wave numbers of electrons and ions. For $Z = 1$ and $T_e = T_i$, one finds $k_{De} = k_{Di}$, so that the contributions of electrons and ions to the Debye screening length are the same.

The static structure factor of the electron-density fluctuations in a two-component plasma near thermodynamic equilibrium ($T_e \simeq T_i$) is given by (e.g., Ichimaru 1962)

$$S(k) = \frac{k^2 + k_{Di}^2}{k^2 + k_{De}^2 + k_{Di}^2} . \tag{4.17}$$

The derivation of Eq. (4.17) is identical in principle to Eq. (4.14). [See Appendix 7 for the derivation.] Assuming $k_{De} = k_{Di}$, one finds $S(k) = 1/2$ in the long-wavelength domain satisfying $k^2 \ll k_{De}^2$. It is noteworthy that this value differs remarkably from that obtained from Eq. (4.14). This magnitude 1/2 stems from the contribution of those electrons residing in the screening clouds in the individual dressed ions. Since the screening effects of electrons and ions are the same, the "dress" of an ion is a 50-50 mixture of electrons and other ions.

We have thus arrived at a complete explanation of both the total strength and the spectral distribution of the scattered signals in Bowles' experiment. Those electrons which form the correlated screening clouds of ions carry the total cross-section as much as a half of the Thomson scattering cross-section of individual electrons even in the long-wavelength domain, $k^2 \ll k_{De}^2$. This one half cannot be distinguished from unity within the experimental accuracies, because for one thing a substantial uncertainty is still involved in the numerical estimate of the electron density in the ionosphere. The dynamic behavior of those electron components is determined by the thermal motion of the ions; hence the spectral spread in the scattered wave would be almost the same as that of the transmitted wave.

## 4.5 Experiments by Pineo, Kraft, and Briscoe

Although one cannot arbitrarily change the plasma temperature in the ionospheric back-scattering experiment, it is possible to increase the frequency spread of the scattered waves by an increase of the frequency of the transmitted radar wave. In Bowles' experiment, the expected spread (1.9 kHz) was smaller than the receiver handwidth (9 kHz), so that the spectral shape of the scattered wave could not be measured.

In 1960, Pineo, Kraft and Briscoe (1960) performed an ionospheric back-scattering experiment in which the transmitter frequency was raised to 440 MHz, almost an order of magnitude greater than that used by Bowles. The corresponding wave number also increases, but one can easily confirm that the condition $k^2 \ll k_{De}{}^2$ is still satisfied. Their experimental parameters are the following: Pulse width 500 μs, pulse repetition rate 50 $s^{-1}$, the peak power 1.6 MW, and the receiver bandwidth 2.3, 11.0, or 38.4 kHz.

As in Bowles' experiment, the total strength of the observed scattered signal was in agreement with that expected from a combination of Thomson scattering cross-section and the estimated electron density. A profile of the ionospheric electron density corresponding to Fig. 4.3 was obtained.

With the increased transmitter frequency in this experiment, it became possible to measure the spectral distribution of the receiver signal as well. Figure 4.5 shows a typical result of the frequency spectrum observed in this experiment. We note that the frequency spread observed is now larger by almost an order of magnitude than that expected in Bowles' experiment. In fact, a theoretical calculation of the frequency spectrum of electron-density fluctuations in a two-component plasma shows an excellent agreement with the data shown in Fig. 4.5. Consequently, these experiments offer a quantitative confirmation of the notion that the spectral characteristics in the ionospheric radar back-scattering experiments are produced by those electron components which are correlated with the ion-density fluctuations in the ionosphere.

## 4.6 Ion-Acoustic Wave

Let us examine in a little more detail the shape of the

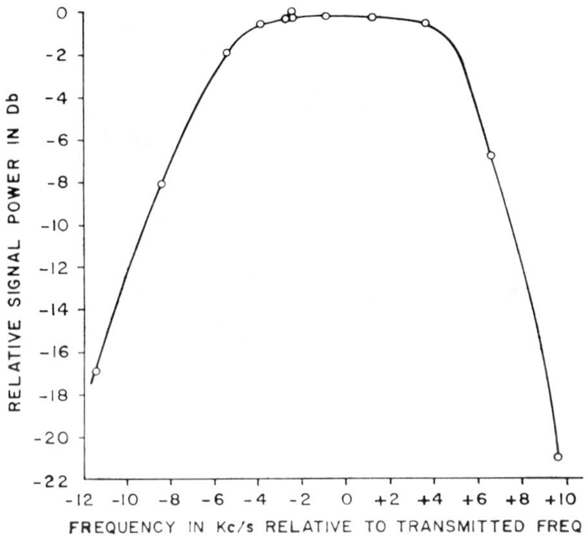


**Figure 4.5.** Spectral distribution of the back-scattered wave from the ionosphere (approximately 300 km in altitude) in the experiment of Pineo et al. (1960). Measured on March 18, 1960. ©American Geophysical Union.

spectral distribution shown in Fig. 4.5. The spectral distribution, plotted in the logarithmic scale along the ordinate, apparently differs from the Gaussian distribution of Fig. 2.5, with shoulders developed in the vicinity of ± 5 kHz.

In terms of the interpretation based on the dressed-particle concept, the frequency spectrum of Fig. 4.5 should reflect the dynamic characters of the ion-density fluctuations. The relevant parameters ($k^2 \ll k_{De}{}^2 \simeq k_{Di}{}^2$) obviously indicate that the collective behavior is essential also for the ions. What is then the characteristic oscillation of the ion system in a two-component plasma, which would correspond to the plasma oscillation for the electron plasma?

We first note in this connection the plasma frequency associated with the ion system,

$$\omega_{pi} = [\frac{4\pi n_i (Ze)^2}{m_i}]^{1/2} \quad . \tag{4.18}$$

As the derivation of Eq. (2.31) clearly indicates, the effective interaction between the ions must be given by the ordinary Coulomb potential $(Ze)^2/r$, in order that Eq. (4.18) should actually correspond to the characteristic frequency of the ion system. The velocity of ionic thermal motion, however, is much smaller than that of electrons owing to the substantial difference between their masses. Consequently, as far as the electron system is concerned, the bulk of ions may be regarded as almost stationary. The electrons can easily respond and relax to the slow motion of ions; the adiabatic approximation is thus applicable. To estimate the effective potential between ions, one can then make a direct use of the analysis carried out in Section 2.2; the inter-ionic effective potential is screened adiabatically by the electrons. Such an effective potential is conveniently expressed in terms of the Fourier components through a replacement

$$\frac{4\pi Ze}{k^2} \rightarrow \frac{4\pi Ze}{k^2 + k_{De}^2} \tag{4.19}$$

accounting for the screening action of the electrons.

With the aid of this replacement, we may proceed to analyze the dynamic behavior of the ion-density fluctuations through the method used in Section 2.5. In place of Eq. (2.29) we now have

$$\dot{\vec{v}}_j = - i \frac{4\pi(Ze)^2}{m_i} \sum_{\vec{q}}{}' \frac{\vec{q}}{q^2 + k_{De}{}^2} \rho_{\vec{q}} \exp(i\vec{q} \cdot \vec{r}_j) \ . \tag{4.20}$$

The characteristic frequency of ions in a two-component plasma is then given by

$$\omega_k = [\frac{4\pi n_i (Ze)^2}{m_i} \frac{k^2}{k^2 + k_{De}{}^2}]^{1/2} \tag{4.21}$$

instead of Eq. (2.32). A derivation of this dispersion relation on the basis of hydrodynamic description will be given later in Section 9.3.*

In the long-wavelength domain $k^2 \ll k_{De}{}^2$, Eq. (4.21) turns into a dispersion relation of acoustic type,

$$\omega_k = s\ k. \tag{4.22}$$

This mode of oscillation is thus called the **ion-acoustic wave**; its sound velocity s is given by

$$s = (Z \frac{k_B T_e}{m_i})^{1/2} = 9.09 \times 10^6\ Z^{1/2} [\frac{m_i}{1.673 \times 10^{-24} g}]^{-1/2} \times [\frac{T_e}{10^6 K}]^{1/2} \quad (\text{cm/s}) \tag{4.23}$$

with the aid of Eqs. (4.16) and (4.21).

---

* Equations (4.21) and (4.25) below can also be obtained from a solution of Eq. (6.15) with Eq. (A7.18), where an appropriate use of the expansions, Eqs. (A5.26) and (A5.28) is to be made.

The amplitude of an ion-acoustic wave may decay or grow through a resonant interaction with charged particles by the mechanism as considered in Section 2.7. The rate of decay or growth is likewise proportional to the gradient of the velocity distribution function of particles at the phase velocity of the ion-acoustic wave [cf. Eq. (2.41) for an OCP case]. When $T_e \simeq T_i$, the sound velocity **s** takes on a value almost the same as the average thermal velocity of the ions, so that the ion-acoustic wave suffers a severe Landau damping due to the ions. In these circumstances, the ion-acoustic wave is so short-lived that it cannot be meaningfully regarded as an oscillatory mode.

For a plasma* with $T_e \gg T_i$, on the other hand, the condition

$$(k_B T_e/m_e)^{1/2} \gg s \gg (k_B T_i/m_i)^{1/2} \tag{4.24}$$

can be well satisfied, because we may usually assume $m_i \gg m_e$ as well; the ion-acoustic wave can avoid a strong Landau damping either from the electrons or from the ions. The rate

---

* A plasma satisfying such a condition can be realized e.g., in an electric-discharge tube. Because $m_e \ll m_i$, the relaxation time $\tau_{ei}$ for energy transfer between electrons and ions takes on a value much greater than $\tau_{ee}$ between electrons or $\tau_{ii}$ between ions. Consequently the equilibration between electrons and ions is by far a slower process than Maxwellization in the electron system or in the ion system. The magnitudes of those relaxation times are relatively

$$\tau_{ee} : \tau_{ii} : \tau_{ei} \sim 1 : (m_i/m_e)^{1/2} : (m_i/m_e) .$$

In the electric discharge the electrons are heated by the applied electric field and attain a high temperature, while the ions remain at around an ambient temperature because of the efficient energy-loss processes through collisions with other atoms and molecules.

of Landau damping $\gamma_k$ of the ion-acoustic wave in this case takes the form (e.g., Ichimaru 1962)

$$\frac{\gamma_k}{\omega_k} = \left(\frac{\pi}{8} Z \frac{m_e}{m_i}\right)^{1/2} \qquad (4.25)$$

in the long-wavelength domain; we thus find $\gamma_k \ll \omega_k$, so that the ion-acoustic wave represents a characteristic oscillation with small damping.

Finally on the basis of the foregoing consideration, let us review the experiments of Pineo et al. For a hydrogen plasma with $T_e \simeq 1500$ K, we find $s \simeq 3.5 \times 10^5$ cm/s from Eq. (4.23). Since $k = 2k_1 = 0.18$ cm$^{-1}$, we confirm $k^2 \ll k_{De}{}^2$ in view of the $k_D$ value computed in Section 4.3; the frequency of the ion-acoustic wave is thus approximately 10 kHz. It is said that the oxygen ions are a main constituent of the ionospheric plasma; in this case the estimated value of the ion-acoustic frequency takes on a value smaller than that. At any rate those ion-acoustic waves suffer severe damping because $T_e \simeq T_i$ in the ionosphere; independent peak structures such as Fig. 2.4 cannot be produced. Rather the ion-acoustic waves act to add weak and broad spectral components, which may result in the shoulders observed in Fig. 4.5. In summary, we may interpret the spectral shoulders in Fig. 4.5 as an indication of the existence of the ion-acoustic wave as a collective motion in a two-component plasma.

## Chapter 5

# Magnetic Confinement

A charged particle in a magnetic field performs a spiral motion along a magnetic line of force; its motion across the magnetic field is, in a sense, suppressed. Associated with the periodic cyclotron motion, one finds an **adiabatic invariant** arising from the action integral over its period, that is the orbital **magnetic moment** of the charged particle. Plasmas confined in magnetic fields are thus characterized by interesting new features such as various **drift motions**; those features in turn affect the efficiency of the confinement. In this chapter we shall study some of those elementary aspects on the **magnetic confinement** of a plasma.

## 5.1 Spiral Motion, Magnetic Moment, and Drifts

The equation of motion for a particle with mass M and electric charge Ze in a constant uniform magnetic field $\vec{B}$ is written as

$$M\dot{\vec{v}} = \frac{Ze}{c} (\vec{v} \times \vec{B}) \quad . \tag{5.1}$$

The velocity $\vec{v}$ and the kinetic energy, $w = (M/2)v^2$, of the particle are decomposed into two components, parallel and perpendicular to the magnetic field:

$$\vec{v} = \vec{v}_{\parallel} + \vec{v}_{\perp} ,$$

$$w = w_{\parallel} + w_{\perp} , \tag{5.2}$$

where

$$w_{\parallel} = \frac{M}{2} v_{\parallel}^2 , \qquad w_{\perp} = \frac{M}{2} v_{\perp}^2 . \tag{5.3}$$

With the aid of Eq. (5.1), one proves that both $w$ and $w_{\parallel}$ are kept constant in time; hence $w_{\perp}$ remains constant as well.

The solution to Eq. (5.1) yields a spiral motion of the charged particle along a magnetic line of force. The **guiding center** which is defined as the center of such a cyclotron motion moves along the magnetic line of force. In terms of the instantaneous position $\vec{r}$ and velocity $\vec{v}$ of the particle, the location $\vec{r}_g$ of the guiding center is given by

$$\vec{r}_g = \vec{r} + \frac{Mc}{ZeB^2} \vec{v} \times \vec{B} . \tag{5.4}$$

The radius $r_L$ of the cyclotron motion, or the **Larmor radius,** is the magnitude of the last term in Eq. (5.4), that is,

$$r_L = \frac{Mv_{\perp}c}{|Z|eB} . \tag{5.5}$$

The angular frequency of gyration is the cyclotron frequency $\omega_H$ expressed as Eq. (4.11). The direction of cyclotron motion depends on the sign of $Z$.

The magnetic moment associated with the cyclotron motion of the charged particle is calculated as

$$\mu_H = \frac{1}{c}\,(\text{current}) \cdot (\text{area})$$

$$= \frac{1}{c}\left(\frac{|Z|e\omega_H}{2\pi}\right)\pi\left(\frac{Mv_\perp c}{|Z|eB}\right)^2 = \frac{w_\perp}{B} \,. \tag{5.6}$$

With inclusion of its direction we write

$$\vec{\mu}_H = -\frac{w_\perp}{B^2}\,\vec{B} \,, \tag{5.7}$$

reflecting the **orbital diamagnetism** of a charged particle. We then calculate the magnetization $\vec{M}$ and the associated magnetization current $\vec{j}_m$ due to the orbital magnetic moment of the charged particles with number density n as

$$\vec{j}_m = c\nabla \times \vec{M} = c\nabla \times (n\vec{\mu}_H) \,. \tag{5.8}$$

We thus find the **diamagnetic drift** given by

$$\vec{v}_D = \frac{\vec{j}_m}{Zen} = -\frac{ck_BT}{ZeB^2}\left(\frac{\nabla n}{n} + \frac{\nabla T}{T}\right) \times \vec{B} \,, \tag{5.9}$$

where a substitution, $nw_\perp = nk_BT$, is made on account of the Maxwellian average with respect to Eq. (1.2).

The diamagnetic drift arises as a consequence of interplay between a spatial inhomogeneity and the finiteness of the Larmor radius. When a density gradient exists, the contributions in particle current from the cyclotron orbits in the high-density side exceed those coming from the low-density side; a net current flow appears as illustrated in Fig. 5.1a. Similarly, when a temperature gradient is involved, the difference between the average velocities in the high- and low-

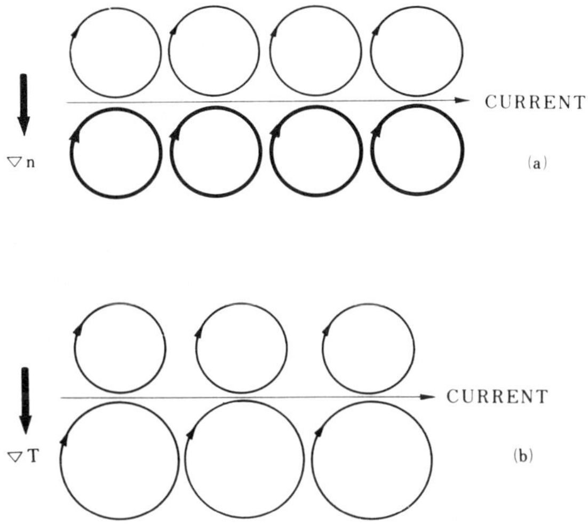


**Figure 5.1.** Diamagnetic currents. The line thickness indicates the number of particles contained in each orbit.

temperature domains accounts for the appearance of current flow, as shown in Fig. 5.1b.

When an external force field $\vec{F}$ is applied in a direction perpendicular to the magnetic field $\vec{B}$, a drift motion of charged particles is induced in the direction perpendicular to both $\vec{F}$ and $\vec{B}$. The equation of motion for a charged particle reads

$$M\dot{\vec{v}} = \vec{F} + \frac{Ze}{c}(\vec{v} \times \vec{B}) \ . \tag{5.10}$$

Defining

$$\vec{v}_F = \frac{c}{ZeB^2} \vec{F} \times \vec{B} \ , \tag{5.11}$$

we substitute $\vec{v} = \vec{v}_F + \vec{v}_1$ in Eq. (5.10), to find

$$M\dot{\vec{v}}_1 = \frac{Ze}{c}(\vec{v}_1 \times \vec{B}) ,$$

an equation identical to Eq. (5.1). Equation (5.10) thus describes a superposition of a uniform drift motion $\vec{v}_F$ and a simple spiral motion $\vec{v}_1$.

When the force under discussion arises from the electric field $\vec{E}$, we substitute $\vec{F} = Ze\vec{E}$ in Eq. (5.11) to obtain the **E × B drift,**

$$\vec{v}_E = c\,\frac{\vec{E} \times \vec{B}}{B^2} . \tag{5.12}$$

When the gravitational acceleration $\vec{g}$ is considered, we have

$$\vec{v}_g = \frac{Mc}{ZeB^2}\,\vec{g} \times \vec{B} , \tag{5.13}$$

the **gravitational drift.**

The physical origin of such a drift is easily understood. Consider, for example, the case of E × B drift as illustrated in Fig. 5.2. A charged particle gyrates in a direction

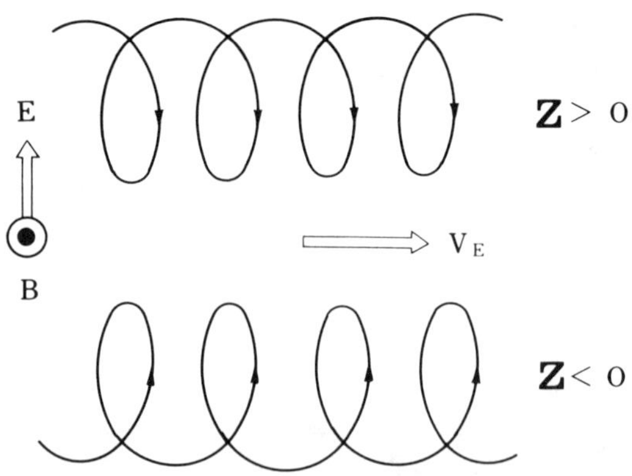


**Figure 5.2.** The E × B drift.

depending on the sign of Z with a radius proportional to $v_\perp$ as given by Eq. (5.5). It is also accelerated by the electric field. On the side where the kinetic energy of the particle is raised by the electric field, the particle orbit assumes a larger radius of curvature; on the other side, where the particle energy is lowered, the particle moves with a smaller radius of curvature. As a result, the drift motion as shown in Fig. 5.2 appears. If the sign of Z is changed, no change in $\vec{\mathbf{v}}_E$ takes place because both the direction of acceleration and the direction of gyration are inverted. In a plasma where the densities of the positive and negative mobile charges compensate each other on the average, the E × B drift does not induce an electric current; it contributes only to the mass motion of the plasma as a whole.

The gravitational drift can be similarly understood. In this case, a change in the sign of charge results in an inversion of the drift direction. The gravitational drift thus acts to induce an electric current even in a macroscopically compensated plasma.

## 5.2 Adiabatic Invariance and Magnetic Mirror

The magnetic moment, Eq. (5.6), is proportional to the action integral associated with the periodic cyclotron motion of the charged particle. It is therefore said to be an **adiabatic invariant**: For an infinitesimally small rate of space-time change in the magnetic field, the magnetic moment is kept invariant.

To illustrate the adiabatic invariance of the magnetic moment, let us consider the effect of $\partial\vec{B}/\partial t$ on the cyclotron motion of a charged particle. The time variation of the magnetic field induces an electric field along the cyclotron orbit according to Eq. (4.2), and thereby acts to change the perpendicular kinetic energy $w_\perp$. We confine ourselves to consideration of only the first-order effects in $\partial\vec{B}/\partial t$. We may thus assume that the particle moves along a closed cyclotron orbit C in calculating the variation of $w_\perp$:

$$\delta w_\perp = Ze \int_C \vec{E} \cdot d\vec{\ell} = Ze \int_A (\nabla \times \vec{E}) \cdot d\vec{A}$$

$$= -\frac{Ze}{c} \int_A \frac{\partial \vec{B}}{\partial t} \cdot d\vec{A} \; .$$

Here the line element $d\vec{\ell}$ along the orbit C should be taken in the direction of particle motion; the direction of the surface element $d\vec{A}$ over the area A bounded by C is determined in accord with the right-hand rule. The product $Zed\vec{A}$ is then pointed oppositely to $\vec{B}$ (cf. the orbital diamagnetism); one thus finds

$$\delta w_\perp = \frac{w_\perp}{B} \left(\frac{2\pi}{\omega_H}\right) \frac{\partial B}{\partial t} \; .$$

Since $2\pi/\omega_H$ is the cyclotron period, $(2\pi/\omega_H)(\partial B/\partial t)$ represents the variation $\delta B$ of the field strength over that period; hence,

$$\delta w_\perp = \frac{w_\perp}{B} \delta B \; . \qquad (5.14)$$

In the light of Eq. (5.6), this implies the adiabatic invariance of the magnetic moment, i.e.,

$$\delta \mu_H = 0 \qquad (5.15)$$

to the first order in $\partial\vec{B}/\partial t$.

One of the important consequences of the adiabatic invariance of the magnetic moment is the notion of the **magnetic-mirror** field constructed by converging magnetic lines of force, as illustrated in Fig. 5.3. As the lines of force converge, the field strength increases. This increasing magnetic field then acts to reflect some of the charged particles impinging on it; hence, the magnetic mirror.

To elucidate the origin of the retarding force, let us analyze the motion of a charged particle in an axisymmetric magnetic field converging in the z direction (Fig. 5.3). The field equation, $\nabla \cdot \vec{B} = 0$, in the cylindrical coordinates reads

$$\frac{1}{r}\frac{\partial}{\partial r}(r\,B_r) + \frac{\partial B_z}{\partial z} = 0 .$$

Since the field under consideration is predominantly in the z direction, we may write $B_z$ as B. In the vicinity of the z axis (i.e., r = 0), the radial component $B_r$ of the field is then expressed as

$$B_r = -\frac{r}{2}\frac{\partial B}{\partial z} . \tag{5.16}$$

The force $F_\parallel$ in the z direction acting on a charged particle whose guiding center is located on the z axis is thus calculated as

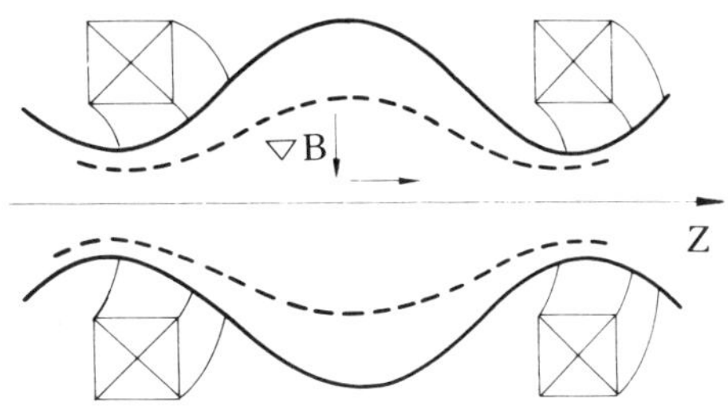


**Figure 5.3.** A magnetic bottle.

$$F_{\parallel} = M\dot{v}_{\parallel} = \frac{|Z|e}{c} v_{\perp} B_r$$

$$= - \frac{|Z|ev_{\perp}r_L}{2c} \frac{\partial B}{\partial z} = -\mu_H \frac{\partial B}{\partial z} . \tag{5.17}$$

As the field strength B increases in the z direction, a charged particle perceives an effective retarding force, Eq. (5.17), proportional to its magnetic moment.

The result of Eq. (5.17) offers in fact another demonstration of the adiabatic invariance of the magnetic moment. To see this, we note on one hand that the total kinetic energy w of the particle is kept constant in a time-independent magnetic field; on the other hand, the parallel component $w_{\parallel}$ is related to the foregoing force by $F_{\parallel} = dw_{\parallel}/dz$. We thus find

$$F_{\parallel} = - \frac{dw_{\perp}}{dz} ,$$

which combined with Eq. (5.17) yields

$$\frac{d\mu_H}{dz} = 0 , \tag{5.18}$$

adiabatic invariance of the magnetic moment to the first order in $\partial B/\partial z$.

One can generalize Eq. (5.17) and show that when a spatial gradient of the field strength exists, a charged particle experiences an effective force given by

$$\vec{F}_G = -\mu_H \nabla B . \tag{5.19}$$

The presence of such an effective force field is again related intimately to the adiabatic invariance of the magnetic moment. A proof of Eq. (5.19) is left to the reader as an exercise (see e.g., Longmire 1963).

## 5.3 Rayleigh-Taylor Instability

The instability of a heavy liquid supported by a light liquid against the gravitational field is a classic problem in hydrodynamics. A similar problem arises in a plasma supported by a magnetic field against gravity or some other force field such as Eq. (5.19). Such instabilities are generally categorized as the **Rayleigh-Taylor instability.** Those involving gravity are in particular called the **gravitational instability.** The Rayleigh-Taylor instability is a typical magnetohydrodynamic instability involving macroscopic mass motion of plasmas.

Consider first the case of the gravitational instability as illustrated in Fig. 5.4a. The plasma, occupying the region above the plane x = 0, is sustained by the magnetic field in

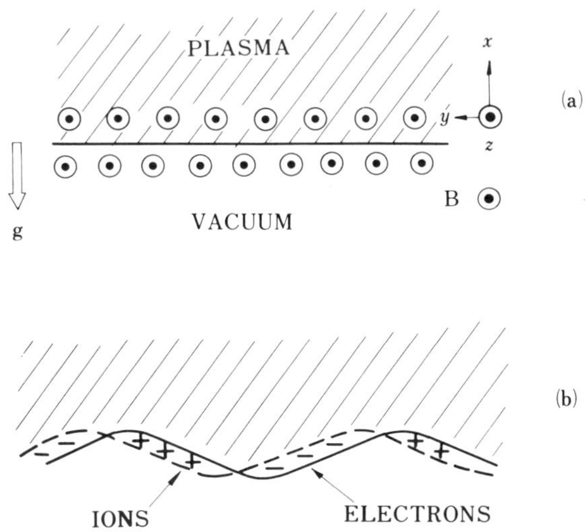


**Figure 5.4.** Gravitational instability.

the z direction against the gravitational field pointed downward in the x direction. If the magnetic field is strong enough, the effect of the gravitational field is only to induce drift motion of the charged particles in the y direction. The plasma could conceivably float above vacuum supported by the magnetic field. The question remains if such a configuration is indeed a stable one.

To examine such a question, we apply a small perturbation

$$\Delta x = \xi \sin ky \tag{5.20}$$

to the boundary plane x = 0 of the macroscopically compensated plasma consisting of the electrons and ions. The magnetic field and the gravitational acceleration induce drift motion of charged particles in the y direction as given by Eq. (5.13). Since the direction of this drift depends on the sign of the charge, a space-charge accumulation takes place on the perturbed boundary, as shown in Fig. 5.4b. The surface charge then produces an electric field, which in turn induces a mass motion of the plasma as a whole in accord with Eq. (5.12). A close examination reveals that the direction of this mass motion is to increase the magnitude $\xi$ of Eq. (5.20); an instability is thus indicated. The plasma cannot be supported against the gravitational field in a stable manner by the magnetic field alone.

If on the other hand the plasma is placed in the region $x < 0$ in Fig. 5.4a rather than in the region $x > 0$, then the induced $E \times B$ drift acts to reduce the displacement $\xi$. The system is therefore stable, as we might naturally expect.

The driving force of the gravitational instability discussed above was provided essentially by the electric field produced by the charge separation at the boundary. This charge separation in turn was brought about by the gravitational field, which is a charge-insensitive force. We can thus develop a similar line of argument to derive a stability criterion, whenever a charge-insensitive force field is involved in a direction perpendicular to the magnetic field. Hence, for a plasma confined in a magnetic field, if there is a force acting on the charged particles, regardless of their signs, in the directions from the plasma to the vacuum region, then the system is found to be unstable with respect to the Rayleigh-Taylor mode. If the force is directed in the opposite directions, the system may be stable as far as the Rayleigh-Taylor mode is concerned.

A most important example of such a charge-insensitive field arises when the magnetic lines of force that confine the plasma have finite radii of curvature. Since the radius vector $\vec{R}$ of the curvature is given by

$$\frac{\vec{R}}{R^2} = - \left(\frac{\vec{B}}{B} \cdot \nabla\right)\frac{\vec{B}}{B} , \tag{5.21}$$

a charged particle moving along the magnetic field experiences a centrifugal force

$$\vec{F}_R = M\, v_\parallel^2 \frac{\vec{R}}{R^2} = 2w_\parallel \frac{\vec{R}}{R^2} . \tag{5.22}$$

The curvature also involves the field gradient in the directions perpendicular to the magnetic lines of force. From a vacuum field relationship $\nabla \times \vec{B} = 0$, which we can reasonably assume for a plasma-confining system, we find

$$\frac{\nabla_\perp B}{B} = - \frac{\vec{R}}{R^2} . \tag{5.23}$$

Substitution of Eq. (5.23) in Eq. (5.19) yields an effective force in the perpendicular directions,

$$\vec{F}_G = w_\perp \frac{\vec{R}}{R^2} \,. \tag{5.24}$$

Collecting the two forces, Eqs. (5.22) and (5.24), we may apply the result of the foregoing gravitational-instability analysis to the present case through a replacement

$$\vec{g} \rightarrow \frac{w_\perp + 2w_\parallel}{M} \frac{\vec{R}}{R^2} = -\left(v_\parallel^{\,2} + \frac{v_\perp^{\,2}}{2}\right)\frac{\nabla_\perp B}{B} \,. \tag{5.25}$$

From the criterion of the Rayleigh-Taylor instability, we then find that a system in which the magnetic field confines the plasma in a convex shape (as in Fig. 5.5a) is unstable because $\vec{R}$ in this case is directed from the plasma to vacuum; on the contrary a system like Fig. 5.5b is stable.

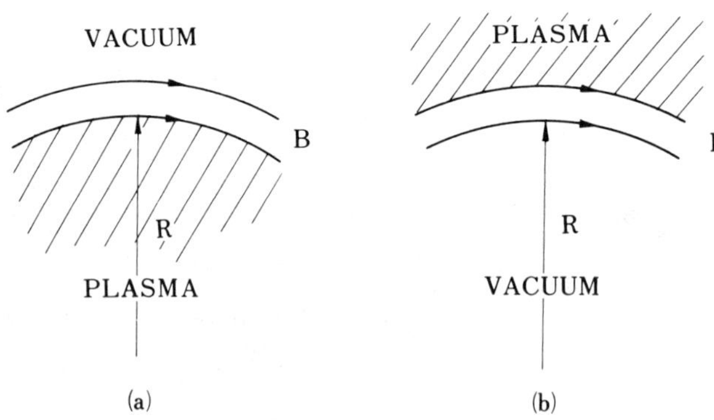


**Figure 5.5.** Stability of magnetic confinement.

## 5.4 Toroidal Confinement

When a plasma is confined in a toroidal magnetic field (see Fig. 5.6), an analogous consideration applies. An acceleration as given by Eq. (5.25) is exerted to the plasma in the direction of the major radius. This acceleration induces an electric current in the direction perpendicular to both the major radius and the axis of the toroidal magnetic field, through the mechanism exemplified by Eq. (5.13). The electric current then acts to polarize the plasma in that direction and a space-charge accumulation takes place in the vicinity of the toroidal wall. The electric field produced by the accumulated space charge drives the entire plasma in the direction of Eq. (5.25), owing to the resultant E × B drift of Eq. (5.12). A plasma confined in a simple toroidal magnetic field is thus magnetohydrodynamically unstable.

A way to avoid such an instability is to impose twisting or a **rotational transform** of the magnetic lines of force around the magnetic axis. The space charges accumulated near the toroidal wall are thereby neutralized as they travel along the twisted magnetic lines of force.

The angle of such a rotational transform is given by

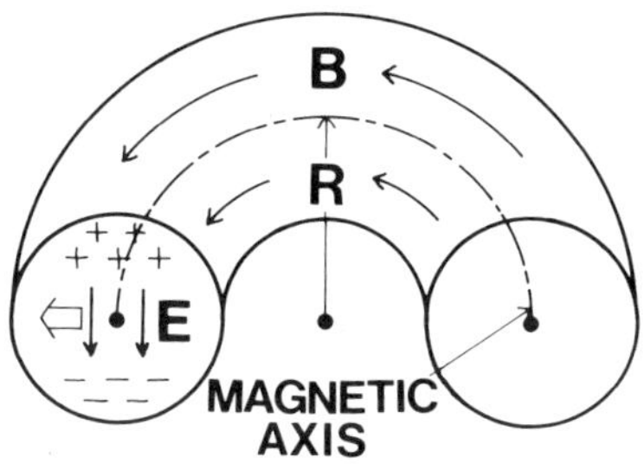


**Figure 5.6.** Toroidal geometry (a cross-sectional view).

$$\iota(r) = \frac{2\pi R B_p(r)}{r\ B_t} \tag{5.26}$$

where R is the major radius, r denotes the distance from the magnetic axis, $B_p$ and $B_t$ refer to the poloidal and toroidal components of the magnetic field. The **safety factor** of the toroidal confinement is then defined as

$$q(r) = 2\pi/\iota(r)\ . \tag{5.27}$$

The magnetohydrodynamic stability of a toroidally confined plasma is analyzed in terms of such a safety factor (e.g., Miyamoto 1980).

Those toroidal surfaces formed by the magnetic lines of force are called the **magnetic surfaces.** A magnetic surface on which the safety factor is given by a rational number

$$q(r) = n/m \tag{5.28}$$

with small (less than 4, say) integral numbers m and n is referred to as a **rational surface.** On such a rational surface, a perturbation propagating along the toroidal axis will resonate with itself after n revolutions. The rational surfaces thus play a special role in the analysis of the evolution of fluctuations in a toroidal system.

Starting with simple concepts as outlined above, one proceeds to develop complex theories of magnetic confinement for fusion reactor systems (e.g., Ribe 1975). A review on ideal magnetohydrodynamic theory of magnetic fusion systems has been given by Freidberg (1982).

## 5.5 Minimum-B Geometry

Equation (5.19) describes the effective force acting upon a charged particle in an inhomogeneous magnetic field when the adiabatic invariance of the magnetic moment is ascertained. We note in this equation that the field strength B plays the part of an effective potential field for the particle. Based on such an observation, we are led to regard the following two features as an essence of designing a magnetohydrodynamically stable field configuration for plasma confinement: (1) The field is nowhere zero; (2) the field strength increases outward from the center toward the periphery of the system. A field configuration satisfying those properties is known as a **minimum-B geometry.**

Property 1 is necessary to maintain the meaning, and thus the adiabatic invariance, of the magnetic moment throughout the system. Property 2 means that B increases outward from a point, or a closed curve, so that surfaces on which B is constant, the **magnetic isobars**, are closed and nested about this point or curve. A contour plot of B therefore gives the shape of a three-dimensional **magnetic well.** It should be noted that the isobars are not the same as the surfaces constructed by the magnetic lines of force, although they may coincide in a few simple geometries.

A best-known example of the minimum-B geometry is the hybrid configuration of the mirror and cusp fields first constructed by Ioffe and his co-workers (Gott, Ioffe, and Telkovsky 1962; Ioffe 1965). A simple mirror field like Fig. 5.3 increases in strength outward along the z axis from the center plane but decreases radially. To obtain Property 2 of a minimum-B configuration we may superpose a multipole field produced by an even number of straight rods laid parallel to the z axis (see Fig. 5.7). Odd and even rods carry current in alternate directions to produce a series of cusped fields (see Fig. 5.7b); these fields obviously increase radially. A minimum-B geometry is thus realized when the rod current exceeds a certain critical value.

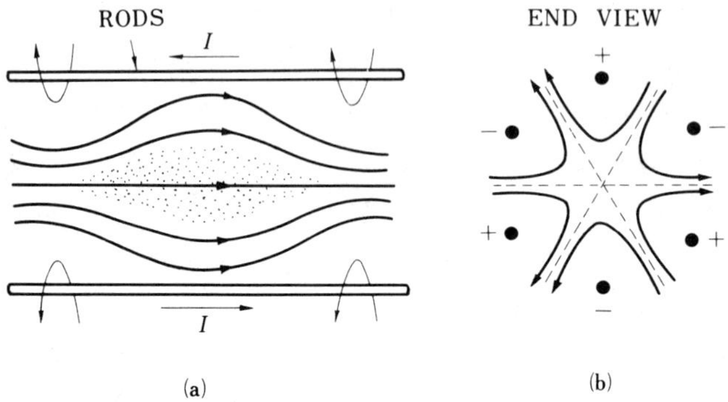


**Figure 5.7.** Mirror-cusp field.

Increasing the current through this critical value, therefore, offers a simple and unambiguous experimental test for the stability properties of the hybrid configuration. As illustrated in Fig. 5.8, such experiments by Ioffe et al. have indeed shown an abrupt increase of the plasma lifetime by more than an order of magnitude when the strength of the cusp fields reaches a certain value, a clear demonstration of the stability achieved by the minimum-B principle.

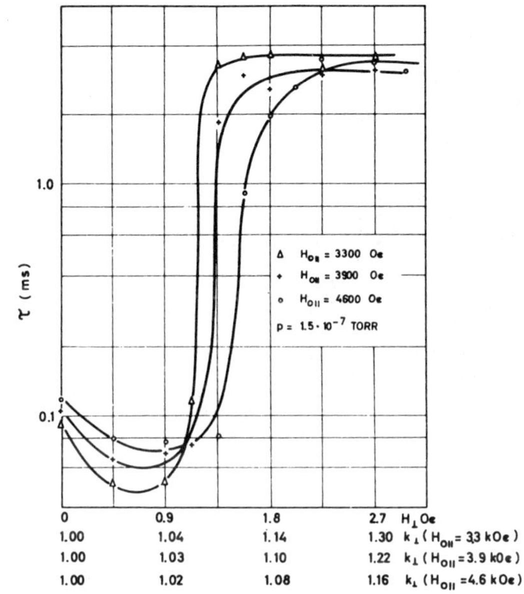


**Figure 5.8.** Dependence of the plasma life-time $\tau$ on the stabilizing field strength. $H_{011}$ is the mirror field at the center; $H_\perp$ is the cusp field close to the wall of the vacuum chamber; the transverse mirror ratio $k_\perp$ is defined by

$$k_\perp = \frac{\left(H_{011}^{2} + H_\perp^{2}\right)^{1/2}}{H_{011}} \quad .$$

## Chapter 6

# Electron Plasmas in Metals

The conduction electrons in metal constitute a typical example of a strongly coupled degenerate plasma. As we examined in Section 1.2, the Coulomb coupling constant $r_s$ for such electrons in metal takes on a value between 2 and 6. It thus makes an interesting physical system in that correlation effects produced by Coulomb interaction together with the quantum-mechanical interference effects between electrons affect the system properties in a fundamental way. To elucidate the properties of such a quantum plasma, various theoretical schemes have been advanced; starting with the random-phase-approximation appropriate to a weakly coupled electron system, and proceeding with inclusion of strong coupling effects through local-field corrections. As a result, we now have a fairly accurate knowledge of the ground state energy and its correlation functions. In this chapter we study those fundamental properties of the electron plasma in metal.

## 6.1 The Fermi Distribution and the Hartree-Fock Exchange Energy

The electron is a fermion with spin quantum number 1/2. Conduction electrons in metal are situated in the periodic electrostatic potential field produced by crystalline ions. One may adopt an approximation in which the ions are replaced by a uniform neutralizing background of positive charges (a jellium model). In other words, the electron system in metal is regarded as a uniform electron gas or liquid. This approximation is sufficiently accurate as far as the plasma properties arising from the Coulomb interaction between electrons are concerned; the effect of the crystalline field may be accounted for in terms of corrections such as the effective mass of an electron.

We consider an electron system in the ground state at T = 0. In such a system the quantum states are occupied by the electrons, one each in a state according to the Pauli principle, successively from the lowest energy level. The number of the states in the wave number space between k and k + dk is $2(2\pi^2)^{-1} k^2 dk$ including the spin degrees of freedom 2. The maximum wave number $k_F$ after placing all the electrons is determined from the condition

$$2 \int_0^{k_F} dk \, \frac{k^2}{2\pi^2} = n \tag{6.1}$$

where n is the number of electrons in a unit volume. Carrying out the integration of Eq. (6.1) and with the aid of Eq. (1.7) we find

$$k_F = \left(3\pi^2 n\right)^{1/3} = 3.63 \; r_s^{-1} \quad [\text{Å}^{-1}] \, . \tag{6.2}$$

As Fig. 6.1 illustrates, the states inside the sphere of radius $k_F$ are all occupied by the electrons. This sphere is the **Fermi sphere** and $k_F$ is referred to as the **Fermi wave**

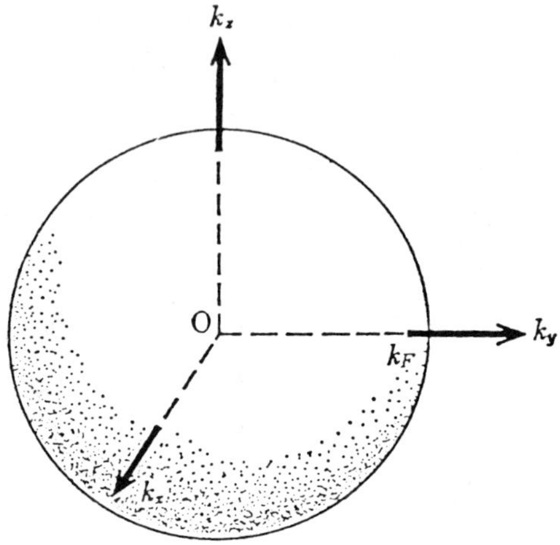


**Figure 6.1.** Fermi sphere.

**number.** The kinetic energy $\hbar^2 k_F^2/2m$ of an electron on the Fermi surface is the Fermi energy $E_F$ given by Eq. (1.4).

The average kinetic energy $E_0$ of an electron in the Fermi sphere is calculated as follows: Since the kinetic energy of an electron in a wave number state k is $\hbar^2 k^2/2m$, we obtain

$$E_0 = \frac{1}{n} \int_0^{k_F} dk \, \frac{\hbar^2 k^2}{2m} \cdot \frac{k^2}{\pi^2} = \frac{3}{5} E_F \qquad (6.3)$$

on referring to Eq. (6.1). Such an energy is customarily measured in units of Rydberg (Ry), the ionization energy of a hydrogen atom in the ground state; it takes on the value

$$1 \text{ Ry} = \frac{me^4}{2\hbar^2} = 13.6 \text{ eV} = 2.18 \times 10^{-11} \text{ erg} . \qquad (6.4)$$

In these units

$$E_0 = \frac{2.21}{r_s^2} \text{ [Ry]} . \tag{6.5}$$

Since the electron is a fermion, a repulsive force is exerted between two electrons with parallel spin on account of the Pauli principle, a manifestation of the exchange interaction. The wave function $\psi(\vec{r}_1, \vec{r}_2)$ of two electrons with parallel spin is antisymmetrical with respect to exchange of the two particle coordinates, $\psi(\vec{r}_1, \vec{r}_2) = - \psi(\vec{r}_2, \vec{r}_1)$, and hence

$$\lim_{\vec{r}_1 \to \vec{r}_2} \psi(\vec{r}_1, \vec{r}_2) = 0 .$$

Consequently the probability of such electrons approaching close to each other is extremely small.

Figure 6.2 shows the radial distribution function, Eq. (3.22), of the electrons calculated by taking account of the correlational effects arising from the exchange interaction (see Appendix 6 for derivation). For those electrons with parallel spin one finds $g(0) = 0$, but the exchange repulsive force does not operate between electrons with antiparallel spin, i.e., $g(0) = 1$; one thus finds $g(0) = 1/2$ on the average.

As we see in Fig. 6.2, the exchange repulsion makes $g(r) < 1$ so that the probability of finding other electrons around a given electron is reduced below the average value unity. As a consequence of such a correlation effect, the interaction energy $E_{int}$ appears. Recalling the definition of the radial distribution function in Section 3.5, we calculate the interaction energy per electron as

$$E_{int} = \frac{n}{2} \int d\vec{r} \, \frac{e^2}{r} \, [g(\vec{r}) - 1] . \tag{6.6}$$

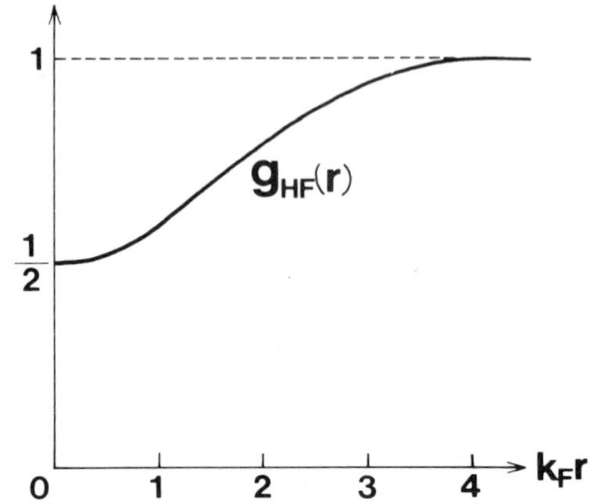


**Figure 6.2.** The radial distribution function in the Hartree-Fock approximation.

The factor 1/2 in Eq. (6.6) is to avoid double counting of the electron interaction; subtraction of 1 from $g(\vec{r})$ in the integrand takes account of the presence of the positive charge background. When electrons are repelled from each other and $g(\vec{r}) < 1$ in the vicinity of $r = 0$, a positive space charge appears from the background. Since an attractive force is exerted between an electron and the positive space charge, the interaction energy takes on a negative value.

Substituting the correlation function of Fig. 6.2 in Eq. (6.6), we obtain

$$E_1 = -\frac{0.916}{r_s}\ [\mathrm{Ry}]\ . \tag{6.7}$$

This is the **Hartree-Fock exchange energy.** At metallic densities ($r_s = 2 - 6$), we find that $|E_1| \simeq E_0$ already.

As may be clear from the foregoing calculation, $E_1$ corresponds to the increment of electrostatic energy arising from the interelectron correlation produced by the Pauli principle. The correlation effect due to the Coulomb interaction has not been taken into account yet. For a strongly coupled plasma with $r_s \gtrsim 1$, we naturally expect a significant contribution coming from the Coulomb-correlation effect. We shall take up these problems starting with the next section.

## 6.2 The Dielectric Response Function

The method of **linear response function** frequently provides a powerful tool in the description of the properties of interacting many-particle systems. Regarding the many-particle system as a "black box," one applies a weak external "disturbance" to probe the system properties. The particles constituting the system are influenced by the disturbance, adjust themselves via interaction, and thereby produce a "response" signal to an external observer. The microscopic properties of the system may be revealed in the way this response is made. When the disturbance is sufficiently weak, we may regard the response as proportional to the disturbance. In these circumstances, the "coefficient of proportionality" between the response and disturbance will reflect the properties of the many-particle system in the **absence** of the external disturbance. This coefficient of proportionality is the linear response function (cf. Appendix 7).

The **dielectric response function** considered in this section is one kind of linear response function, and plays an essential part in the description of plasma properties due to Coulomb interaction. To extract Coulomb-correlational properties, we assume a weak potential field of plane-wave type

$$\phi_{ext}(\vec{\mathbf{r}}, t) = (\phi_{ext})_{\vec{\mathbf{k}}\omega} \exp[i(\vec{\mathbf{k}} \cdot \vec{\mathbf{r}} - \omega t) + \eta t] \qquad (6.8)$$

$$+ \text{(complex conjugate)}$$

as the external disturbance, and consider the charge-density fluctuation

$$Ze\langle\rho_{ind}(\vec{r}, t)\rangle = Ze\langle\rho_{ind}\rangle_{\vec{k}\omega} \exp[i(\vec{k} \cdot \vec{r} - \omega t) + \eta t] \qquad (6.9)$$

$$+ \text{(complex conjugate)}$$

thereby induced in the plasma. In the linear-response formalism, we may take a plane-wave response, Eq. (6.9), corresponding to the plane-wave disturbance, Eq. (6.8). The dielectric response function $\varepsilon(\vec{k}, \omega)$ is then defined in terms of the ratio between those complex amplitudes as

$$\frac{1}{\varepsilon(\vec{k}, \omega)} = 1 + \frac{4\pi Ze}{k^2} \frac{\langle\rho_{ind}\rangle_{\vec{k}\omega}}{(\phi_{ext})_{\vec{k}\omega}} \, . \qquad (6.10)$$

The quantity $\eta$ contained in Eqs. (6.8) and (6.9) is a positive infinitesimal, used here to ensure the requirement for the principle of causality which states that the excitation of the density fluctuation, Eq. (6.9), **follows** the application of the potential of external disturbance.

That $\varepsilon(\vec{k}, \omega)$ defined in Eq. (6.10) has the meaning of a dielectric function can be understood in the following way: The effective potential $\phi_{\vec{k}\omega}$ in the plasma may be expressed as a summation of the external potential and the potential due to the induced charge,

$$\phi_{\vec{k}\omega} = (\phi_{ext})_{\vec{k}\omega} + \frac{4\pi Ze}{k^2} \langle\rho_{ind}\rangle_{\vec{k}\omega} \, . \qquad (6.11)$$

Elimination of $\langle\rho_{ind}\rangle_{\vec{k}\omega}$ between Eqs. (6.10) and (6.11) yields

$$\phi_{\vec{k}\omega} = \frac{(\phi_{ext})_{\vec{k}\omega}}{\varepsilon(\vec{k}, \omega)} . \tag{6.12}$$

Thus the effective potential in the plasma is the external potential $(\phi_{ext})_{\vec{k}\omega}$ multiplied by $[\varepsilon(\vec{k}, \omega)]^{-1}$. We note that $\nabla\phi_{ext}$ and $\nabla\phi$, respectively, correspond to $-\vec{D}$ and $-\vec{E}$ in the ordinary electromagnetism; hence Eq. (6.12) implies $\varepsilon\vec{E} = \vec{D}$.

We remark that Eq. (2.5) in Section 2.2 is a special case of Eq. (6.11). Adopting Eq. (2.4) as $\phi_{ext}(\vec{r}, t)$ one finds

$$(\phi_{ext})_{\vec{k}\omega} = \frac{4\pi Z_0 e}{k^2} 2\pi\delta(\omega) . \tag{6.13}$$

Equation (2.5) is then obtained by substituting Eq. (6.13) in Eq. (6.11), multiplying both sides by $k^2$, and carrying out the inverse Fourier transformation.

Once the dielectric response function is determined, we may calculate the spectral function of density fluctuations for a plasma in thermodynamic equilibrium, that is, the dynamic structure factor (3.16), with the aid of the **fluctuation-dissipation theorem.** This is one of the fundamental theorems in statistical mechanics. We refer the reader to Appendix 7 for the fluctuation-dissipation theorem and its applications. Here we simply cite the result (A7.15) for the electron system in metal

$$S(\vec{k}, \omega) = -\frac{\hbar k^2}{4\pi^2 e^2} \operatorname{Im} \frac{1}{\varepsilon(\vec{k}, \omega)} \qquad (\omega \geq 0) . \tag{6.14}$$

In Section 2.6, we considered plasma oscillation, a typical example of collective density fluctuations in a plasma. Equation (6.10) implies that a density fluctuation with

$\langle \rho_{ind} \rangle_{\vec{k}\omega} \neq 0$ can be spontaneously excited even without an external potential field, if

$$\varepsilon(\vec{k}, \omega) = 0 \qquad (6.15)$$

is satisfied. In other words, Eq. (6.15) gives the **dispersion relation** between the frequency $\omega$ and the wave number $\vec{k}$ of density fluctuations.

Since $\varepsilon(\vec{k}, \omega)$ is a complex function in general, Eq. (6.15) implies two equations coming from its real and imaginary parts. When k is regarded as a real quantity, it then determines a complex frequency $\omega_k - i\gamma_k$; when $\omega$ is real, a complex wave number $k_r(\omega) + ik_i(\omega)$ is determined. Those mutual relations were mentioned in Section 3.1 in connection with the measurement of propagation characteristics of space-charge waves. If $|\gamma_k/\omega_k| << 1$, a characteristic oscillation with small damping exists and the Dynamic structure factor Eq. (6.14) exhibits sharp peaks as shown in Fig. 2.4.

### 6.3 The Random-Phase Approximation

In the high-density limit ($r_s \to 0$) of a degenerate electron gas, the kinetic energy in the Fermi sphere constitutes the principal contribution to the system energy, as the comparison of Eq. (1.6) illustrates. One can thus ignore the density fluctuations and the correlation effect due to the Coulomb interaction in the high-density limit. When an external disturbance (6.8) is applied to such a weakly coupled plasma, one can likewise neglect the nonlinear coupling between the disturbance and the fluctuations in the plasma. The method of calculating the response function by ignoring the Coulomb correlations in the ground state is thus equivalent to the random-phase approximation (RPA) introduced in Section 2.6, in which nonlinear coupling between density fluctuations is neglected.

The dielectric response function calculated in the RPA is the **Lindhard dielectric function** (Lindhard 1954). Leaving its derivation to Appendix 5, we list the results, Eqs. (A5.16) and (A5.17), in the form

$$\varepsilon(\vec{\mathbf{k}}, \omega) = 1 - \frac{4\pi e^2}{k^2} \chi_0(\vec{\mathbf{k}}, \omega) \tag{6.16}$$

where

$$\chi_0(\vec{\mathbf{k}}, \omega) = \frac{2}{\hbar} \sum_{\vec{\mathbf{p}}} \frac{F(\vec{\mathbf{p}}) - F(\vec{\mathbf{p}} + \hbar\vec{\mathbf{k}})}{\omega - \vec{k} \cdot \vec{p}/m - \hbar k^2/2m + i\eta} \tag{6.17}$$

is the **free-electron polarizability.** The momentum distribution function $F(\vec{\mathbf{p}})$ in Eq. (6.17) takes on values

$$F(\vec{\mathbf{p}}) = \begin{cases} 1 & (p \leq \hbar k_F) \\ 0 & (p > \hbar k_F) \end{cases} \tag{6.18}$$

representing the Fermi distribution of Fig. 6.1.

Let us investigate the solution to the dispersion relation (6.15) for the plasma oscillation predicted by the Lindhard dielectric function (6.16). The electron velocity takes its maximum value, $v_F = \hbar k_F/m$, on the Fermi sphere. Assuming $|\omega/k| \gg v_F$, we expand the factor $(\omega - \vec{\mathbf{k}} \cdot \vec{\mathbf{p}}/m - \hbar k^2/2m)^{-1}$ in Eq. (6.17) in a power series of $\omega^{-1}$; the result is

$$\mathrm{Re}\ \varepsilon(\vec{\mathbf{k}}, \omega) = 1 - \frac{\omega_p^2}{\omega^2} - \frac{8\pi e^2}{m\omega^4} \sum_{\vec{p}} F(\vec{\mathbf{p}}) \left[3\left(\frac{\vec{\mathbf{k}} \cdot \vec{p}}{m}\right)^2 + \frac{\hbar^2 k^4}{4m^2} + \ldots\right] . \tag{6.19}$$

Retaining up to the terms proportional to $k^2$ in the long-wavelength domain ($k^2 \ll k_F^2$), we find

$$\omega_k^2 = \omega_p^2 + \frac{3}{5} v_F^2 k^2 \tag{6.20}$$

as a solution to Re $\varepsilon(\mathbf{k}, \omega) = 0$. This is the RPA dispersion relation of the plasma oscillation in a degenerate electron plasma. This result can be derived directly by an application of Eq. (6.18) to Eq. (2.38).

We next turn to evaluation of the imaginary part of $\varepsilon(\mathbf{k}, \omega)$. The $\vec{\mathbf{p}}$ summation in Eq. (6.17) can be replaced by an integration [cf. (A2.5)]; with the aid of Eq. (A1.6), we find

$$\text{Im } \varepsilon(\vec{\mathbf{k}}, \omega) = -\frac{8\pi^2 e^2}{\hbar k^2} \int \frac{d\vec{\mathbf{p}}}{(2\pi\hbar)^3} F(\vec{\mathbf{p}}) \left[ \delta\left(\omega - \frac{\vec{k} \cdot \vec{p}}{m} + \hbar \frac{k^2}{2m}\right) - \delta\left(\omega - \frac{\vec{k} \cdot \vec{p}}{m} - \hbar \frac{k^2}{2m}\right) \right] . \tag{6.21}$$

Since $|\vec{\mathbf{p}}|/m \lesssim v_F$, the domain for Im $\varepsilon(\vec{\mathbf{k}}, \omega) = 0$ is limited to

$$-k v_F + \hbar \frac{k^2}{2m} < \omega < k v_F + \hbar \frac{k^2}{2m} \tag{6.22}$$

as shown in Fig. 6.3. As the dispersion relation (6.20) enters this domain, the plasma oscillation couples resonantly with electrons and thereby suffers a severe Landau damping.

When Eq. (6.20) is outside the domain (6.22), on the other hand, the plasma oscillation suffers no damping; the spectral width of density fluctuations vanishes. This is a special feature of the RPA calculation, which can be examined through comparison with experiments.

The upper limit $k_c$ of the wave number domain where the plasma oscillation receives no Landau damping can be determined from the intersection between Eq. (6.20) and

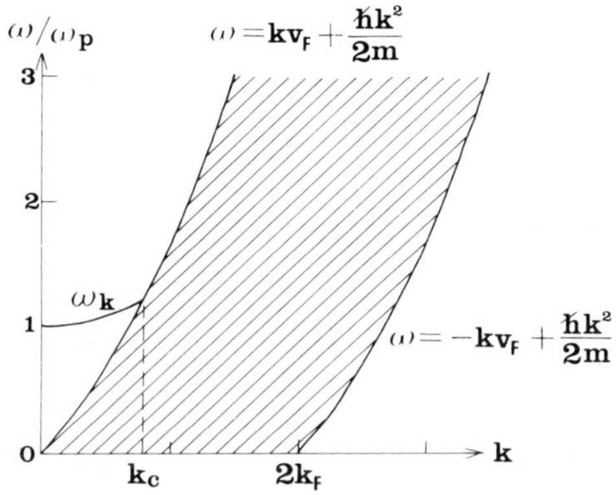


**Figure 6.3.** Plasmon $\omega_k$ and the single pair excitations (hatched area) in the random-phase approximation.

$$\omega = k\ v_F + \hbar\,\frac{k^2}{2m} \qquad (6.23)$$

as Fig. 6.3 illustrates. For simplicity, we approximate Eq. (6.20) as $\omega_p = \omega_p$ and Eq. (6.23) as $\omega = kv_F$, to obtain

$$k_c = \frac{\omega_p}{v_F} = 1.7\ r_s^{-1/2} \quad (\text{Å}^{-1}) \ . \qquad (6.24)$$

This value can likewise be checked through comparison with experiments.

## 6.4 Local-Field Corrections

For a weakly coupled plasma ($r_s \ll 1$ or $\Gamma \ll 1$), we expressed the dielectric response function in the RPA form of Eq. (6.16). This expression, however, has a number of severe difficulties, which we shall now consider.

We begin with the calculation of the radial distribution function of a classical plasma in thermodynamic equilibrium. In this case the Vlasov dielectric function, Eq. (A5.19), is applicable, which is the classical counterpart to Eqs. (6.16) and (6.17) in the RPA. Vlasov expression leads to Eq. (4.14) via Eq. (A7.17); the radial distribution function is then calculated with the aid of Eq. (3.22) as

$$g(\vec{r}) = 1 - \frac{(Ze)^2}{k_B T r} \exp\left(-\frac{r}{\lambda_D}\right) . \tag{6.25}$$

This is the **Debye-Hückel correlation function.** In the short-range domain such that $r < (Ze)^2/k_B T$, Eq. (6.25) would predict $g(\vec{r}) < 0$. This behavior is physically unacceptable, because a probability distribution should not be negative. For the Lindhard dielectric function the case with $r_s = 5$ is depicted in Fig. 6.4 (RPA in the inset); obviously a similar difficulty is found in this case as well.

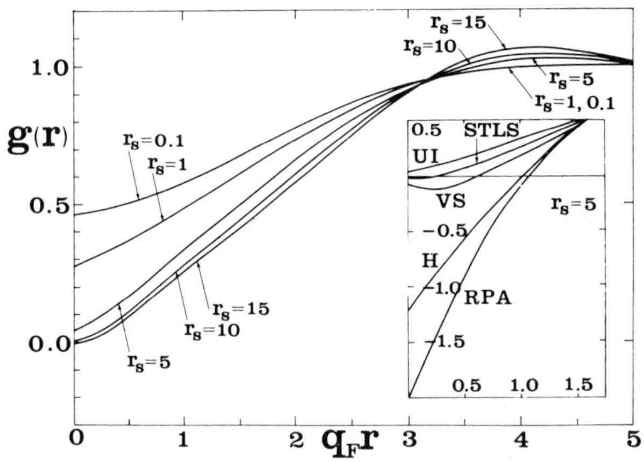


**Figure 6.4.** Radial distribution function g(r) vs. $q_F r$ where $q_F$ is the Fermi wave number. After Utsumi and Ichimaru (1980b). In the inset, the predictions of various theories on the short-range behavior of g(r) are compared at $r_s = 5$. RPA refers to the theory with G(k) = 0, i.e., Eq. (6.16); H with Eq. (6.33); STLS with Eq. (6.39); VS with Eq. (6.41); and UI in Utsumi and Ichimaru (1980b).

The origin of this difficulty may be traced to the fact that the RPA dielectric function does not take account of interparticle correlations in its procedure of calculation. The correlation effects are particularly pronounced at short distances due to interparticle Coulomb repulsion even in a weakly coupled plasma. As a result, Vlasov or Lindhard expression, ignoring correlations altogether, exhibits conspicuous inaccuracy in the prediction of short-range correlations. In the treatment of strongly coupled plasmas such as the electron system in metal, we must squarely face with the correlation effects.

As a theoretical method of formulating such a strong-correlation effect, we introduce the concept of **local-field correction** in this section. In this method the following approximation is adopted in the expression for the dielectric response function.

Let the effective interaction potential $(\phi_{pol})_{\vec{k}\omega}$ produced by the presence of a Fourier component $\langle\rho_{ind}\rangle_{\vec{k}\omega}$ of density fluctuations in the plasma be expressed as

$$Ze(\phi_{pol})_{\vec{k}\omega} = \psi(\vec{k}, \omega)\langle\rho_{ind}\rangle_{\vec{k}\omega} . \tag{6.26}$$

The function $\psi(\vec{k}, \omega)$ introduced here generally differs from the simple Coulomb field $4\pi(Ze)^2/k^2$, as one might anticipate from the presence of screening effect, for example.

We then introduce a density response function $\chi_{sc}(\vec{k}, \omega)$ against the electrostatic potential $(\phi_{ext})_{\vec{k}\omega} + (\phi_{pol})_{\vec{k}\omega}$ in the plasma according to

$$\langle\rho_{ind}\rangle_{\vec{k}\omega} = \chi_{sc}(\vec{k}, \omega)Ze[(\phi_{ext})_{\vec{k}\omega} + (\phi_{pol})_{\vec{k}\omega}] . \tag{6.27}$$

The dielectric response function determined from the definition (6.10) becomes

$$\varepsilon(\vec{k}, \omega) = 1 - \frac{4\pi(Ze)^2}{k^2} \frac{\chi_{sc}(\vec{k}, \omega)}{1 - \{\psi(\vec{k}, \omega) - \frac{4\pi(Ze)^2}{k^2}\}\chi_{sc}(\vec{k}, \omega)} . \tag{6.28}$$

Equation (6.28) in effect amounts to rewriting an unknown function $\varepsilon(\vec{k}, \omega)$ for a strongly coupled plasma in terms of two unknown functions $\psi(\vec{k}, \omega)$ and $\chi_{sc}(\vec{k}, \omega)$. In view of the response relation (6.27), then, the principal correlation effects have been taken into account through Eq. (6.26) in the electrostatic potential $(\phi_{ext})_{\vec{k}\omega} + (\phi_{pol})_{\vec{k}\omega}$; we may thus adopt an approximation in which the free-electron polarizability (6.17) is used for the expression of $\chi_{sc}(\vec{k}, \omega)$. The features of this method, therefore, are the introduction of an effective interaction $\psi(\vec{k}, \omega)$ between microscopic density fluctuations and the consideration of a density response against a total potential field which renormalizes such an effective interaction.

As may be clear from a comparison between Eqs. (6.16) and (6.28), an RPA result, Lindhard or Vlasov expression, is equivalent to taking $\psi(\vec{k}, \omega) = 4\pi(Ze)^2/k^2$ and using the free-electron polarizability $\chi_0(\vec{k}, \omega)$ in place of $\chi_{sc}(\vec{k}, \omega)$. Consequently the change of effective interaction due to interparticle correlations is not taken into account.

To single out the effects of static correlations such as screening, we may ignore the $\omega$ dependence of $\psi(\vec{k}, \omega)$ and replace it as

$$\psi(\vec{k}, \omega) \xrightarrow[(\omega \to 0)]{} \frac{4\pi(Ze)^2}{k^2} [1 - G(\vec{k})] . \tag{6.29}$$

Modifying the RPA in the sense of Eq. (6.29), we write the dielectric function as

$$\varepsilon(\vec{k}, \omega) = 1 - \frac{4\pi(Ze)^2}{k^2} \frac{\chi_0(\vec{k}, \omega)}{1 + G(\vec{k}) \dfrac{4\pi(Ze)^2}{k^2} \chi_0(\vec{k}, \omega)} \tag{6.30}$$

The function $G(\vec{k})$ introduced via Eq. (6.29) describes the extent to which the effective interaction between particles deviates from the simple Coulomb interaction $4\pi(Ze)^2/k^2$ owing to the correlation effect; it is thus called the local-field correction.

The exchange repulsive forces between electrons treated in Section 6.1 can be taken into consideration in the form of the local-field correction as follows. As Fig. 6.2 has demonstrated, electrons with parallel spin cannot approach within a distance of $\sim k_F^{-1}$ because of the Pauli principle. In such a short distance only the Coulomb forces between electrons of antiparallel spin remain; the effective potential becomes half of the ordinary Coulomb potential. Since $r < k_F^{-1}$ corresponds to $k > k_F$ in the wave number space, we may assume

$$G(\vec{k}) \to \frac{1}{2} \qquad (k^2 \gg k_F^2) \tag{6.31}$$

because of Eq. (6.29). In the long-distance domain $r > k_F^{-1}$, on the other hand, the correlation effect is weak and thus the RPA may be applicable with good accuracy; hence we expect

$$G(\vec{k}) \to 0 \qquad (k^2 \ll k_F^2) . \tag{6.32}$$

Hubbard (1958) applied the diagrammatic technique to the theory of electron gas and thereby derived a local-field correction stemming from the exchange interaction

$$G_H(\vec{k}) = \frac{k^2}{2(k^2 + k_F^2)} , \qquad (6.33)$$

which satisfies the conditions (6.31) and (6.32). Nozières and Pines (1958), noting similar features, treated the long- and short-range domains separately and connected those evaluations later in the calculation of the static structure factor (3.20).

In the long-wavelength limit, the local-field correction $G(\vec{k})$ is expressed in a form proportional to $k^2$, so that

$$\lim_{k \to 0} G(\vec{k}) = \gamma_0(r_s)(k/k_F)^2 . \qquad (6.34)$$

The coefficient $\gamma_0(r_s)$ is connected with the thermodynamic compressibility,

$$\kappa = -[V \frac{\partial P}{\partial V}]^{-1} , \qquad (6.35)$$

via the **compressibility sum rule** (e.g., Pines and Nozières 1966); (Ichimaru 1982) as

$$\frac{\kappa_0}{\kappa} = 1 - \frac{4r_s}{\pi} (\frac{4}{9\pi})^{1/3} \gamma_0(r_s) \qquad (6.36)$$

where $\kappa_0 = 3/2E_F n$ is the compressibility of the noninteracting Fermi system.

A thermodynamic consideration related to the compressibility sum rule stated above subsequently revealed that the local-field correction arising from the exchange interactions should behave as

$$G(\vec{k}) \longrightarrow k^2/4k_F^2 \qquad (k^2 \ll k_F^2) \qquad (6.37)$$

in the long-wavelength limit. Geldart and Vosko (1966) thus modified Eq. (6.33) and proposed

$$G_{GV}(\vec{k}) = \frac{k^2}{2(k^2 + 2\,k_F^{\,2})} \,. \tag{6.38}$$

Various proposals have been advanced to take account of the Coulomb correlation effects in the calculation of the dielectric response function. In this connection it has been noted essential that the correlation effects be taken into account in a self-consistent way. Assume for the moment that the function $\psi(\vec{k}, \omega)$ representing the effective interaction has been somehow determined. $\varepsilon(\vec{k}, \omega)$ contains $\psi(\vec{k}, \omega)$ through Eq. (6.28). The dynamic structure factor $S(\vec{k}, \omega)$ describing the density correlations, on one hand, is connected with $\varepsilon(\vec{k}, \omega)$ via the fluctuation-dissipation theorem (6.14). The presence of the density correlations, on the other hand, provides a principal factor in determining the effective interaction. For a self-consistent theory, the $\psi(\vec{k}, \omega)$ originally assumed must agree with the resultant $\psi(\vec{k}, \omega)$ derived as a consequence.

Singwi, Tosi, Land, and Sjölander (1968) proposed such a self-consistent theory, called the STLS scheme. In Eq. (6.28), $\chi_{sc}(\vec{k}, \omega) = \chi_0(\vec{k}, \omega)$ is assumed and the approximation (6.29) is adopted. The local-field correction $G(\vec{k})$ is expressed as a functional of the static structure factor in the following way.

In the absence of interparticle correlations, the force field exerted by the electron density fluctuations is expressed in the Fourier component as

$$- i\vec{k}\,\frac{4\pi e^2}{k^2}\,\langle\rho_{ind}\rangle_{\vec{k}\omega} \,.$$

In the presence of correlations the probability of finding

other electrons at $\vec{r}'$ increases by a factor of $[g(\vec{r}') - 1]$; these electrons produce a potential proportional to $|\vec{r} - \vec{r}'|^{-1}$ at $\vec{r}$. Consequently, the effective force with correlation effects may be evaluated as

$$-\left\{i\vec{k}\,\frac{4\pi e^2}{k^2} + \frac{i}{n}\int \frac{d\vec{q}}{(2\pi)^3}\,\vec{q}\,\frac{4\pi e^2}{q^2}\,[S(\vec{k} - \vec{q}) - 1]\right\}\langle\rho_{ind}\rangle_{\vec{k}\omega}$$

with the aid of the relation (3.22). From the ratio between those two evaluations, Singwi et al. derived an expression

$$G_{STLS}(\vec{k}) = -\frac{1}{n}\int \frac{d\vec{q}}{(2\pi)^3}\,\frac{\vec{k}\cdot\vec{q}}{q^2}\,[S(\vec{k} - \vec{q}) - 1] \tag{6.39}$$

for the local-field correction.

The dielectric response function, Eq. (6.30), is now expressed as a functional of the static structure factor via Eq. (6.39). A combination of this expression with Eq. (6.14) then makes a set of self-consistent integral equations for determination of $S(\vec{k})$ or $G(\vec{k})$, which is to be solved by a method of iteration. The results so obtained were found to represent a significant improvement over the RPA predictions both on the behavior of the short-range correlations (see Fig. 6.4) and on the values of the correlation energy to be considered in the next section.

A number of shortcomings were noted on the STLS scheme at the same time: When the theory was applied to a weakly coupled, classical OCP ($\Gamma \ll 1$), it failed to reproduce the known expansion, Eq. (1.11), of the internal energy density. A second problem on the STLS scheme concerns the compressibility sum rule, mentioned above in connection with Eq. (6.37). The theory exhibited a substantial discrepancy between the values of the compressibility Eq. (6.35) calculated thermodynamically and those obtained from the long-wavelength limit Eq. (6.36) of $G_{STLS}(\vec{k})$.

As may be clear from its derivation, Eq. (6.39) does not take into account the effects of ternary correlations. Ichimaru (1970) adopted the convolution approximation* in the expression for the ternary correlation function and thereby derived a local-field correction,

$$G_{CA}(\vec{k}) = -\frac{1}{n}\int \frac{d\vec{q}}{(2\pi)^3}\,\frac{\vec{k}\cdot\vec{q}}{q^2}\,S(\vec{q})[S(\vec{k}-\vec{q})-1]\;. \tag{6.40}$$

Analytic structures of and numerical solution to the resulting integral equations were subsequently studied by Totsuji and Ichimaru (1973; 1974); it was shown explicitly that Eq. (6.40) reproduces the Abe formula (1.11). The convolution scheme, Eq. (6.40), thus offers a description of static properties in the classical OCP superior to the STLS scheme, Eq. (6.39).

Subsequently, through investigation of the compressibility, Schneider (1971) and Vashishta and Singwi (1972) proposed the form

$$G_{VS}(\vec{k}) = [1 + a_0 n\,\frac{\partial}{\partial n}]G_{STLS}(\vec{k})\;. \tag{6.41}$$

The parameter $a_0$ is to be chosen in such a way that the resulting formulation satisfies the compressibility sum rule; for the classical OCP, $a_0 = 1/2$, and for the degenerate electron gas, $a_0 = 2/3$. This formula, however, fails to reproduce Eq. (1.11) beyond the Debye-Hückel term proportional to $\Gamma^{3/2}$.

Recently, Utsumi and Ichimaru (1980 a, b) advanced a micro-

---

* See Section 10.3 for a detailed discussion.

scopic theory in which the contributions from both the exchange and nonlinear Coulomb correlations are explicitly taken into account in the calculation of the local-field correction. Figure 6.4 exhibits the values of the radial distribution function $g(\vec{r})$ calculated in this theory, together with comparison between various theories. If the positivity of $g(\vec{r})$ may be taken as a criterion for judgment on the accuracy of a theory, we find that a substantial improvement has been achieved in recent theories. A more detailed comparison is found in a recent review (e.g., Ichimaru 1982).

## 6.5 The Ground-State Energy

In the description of macroscopic properties of ordinary physical systems, thermodynamic functions such as the free energy play an important part. The ground-state energy which we are concerned with in this section is one of those quantities for the electron system in metal.

The **ground-state energy** E refers to the energy per electron for the electron system in the ground state at T = 0. This quantity in Rydberg units, Eq. (6.4), is a function of only $r_s$; it is thus written as $E(r_s)$.

The calculation of the ground-state energy is facilitated by the following **ground-state energy theorem:** Let $\Phi_0$ be the ground-state wave function of the n electrons in a unit volume with Hamiltonian H; it then satisfies the Schrödinger equation

$$H \, \Phi_0 = nE \, \Phi_0 \, .$$

Since $\Phi_0$ depends on the strength of Coulomb coupling $e^2$, we write it as $|e^2\rangle$ and impose the normalization condition

$$\langle e^2 | e^2 \rangle = 1 \, .$$

To calculate $nE(r_s) = \langle e^2 | H | e^2 \rangle$, we differentiate it with respect to $e^2$:

$$\frac{d(nE)}{de^2} = \langle e^2 | \frac{dH}{de^2} | e^2 \rangle + nE \frac{d}{de^2} \langle e^2 | e^2 \rangle .$$

The last term vanishes because of the normalization condition. Since in the light of Eq. (6.6)

$$\langle e^2 | \frac{dH}{de^2} | e^2 \rangle = \frac{nE_{int}(e^2)}{e^2}$$

and $nE = nE_0$ at $e^2 = 0$, we obtain the relation

$$E = E_0 + \int_0^{e^2} dx \frac{E_{int}(x)}{x} , \tag{6.42}$$

which is the ground-state energy theorem for an electron gas.

When the interaction between electrons is neglected altogether, the ground-state energy is thus given by Eq. (6.5). When the exchange interaction between electrons is taken into account, application of Eq. (6.7) to Eq. (6.42) leads to

$$E_{HF}(r_s) = \frac{2.21}{r_s^2} - \frac{0.916}{r_s} \quad [Ry] . \tag{6.43}$$

This is the **ground-state energy in the Hartree-Fock approximation.**

The radial distribution function $g(\vec{r})$ between electrons, when the Coulomb interaction is taken into account, differs from the radial distribution function in the Hartree-Fock approximation (see Fig. 6.2), as Fig. 6.4 illustrates. This difference produces the **correlation energy**, $E_c(r_s)$, in the estimate of the ground-state energy through Eq. (6.6). The definition of the correlation energy is thus the difference of the ground-state energy $E(r_s)$ with Coulomb correlations from the ground-state energy in the Hartree-Fock approximation,

$$E_c(r_s) = E(r_s) - E_{HF}(r_s) \ . \tag{6.44}$$

Gell-Mann and Brueckner (1957) started with a diagrammatic theory equivalent to the RPA dielectric function (6.16) and proceeded to calculate the ground-state energy through the process of Eqs. (6.14), (3.19), (3.22), and (6.6). In addition they evaluated numerically the contribution of the second-order exchange terms to the ground-state energy, which is not included in Fig. 6.2. They thereby determined the first few terms of the high-density expansion ($r_s$ expansion) of the ground-state energy; the result is

$$E_{GB}(r_s) = \frac{2.21}{r_s^{\,2}} - \frac{0.916}{r_s} + 0.062 \ \ln r_s - 0.096 \quad [\mathrm{Ry}] \ , \tag{6.45}$$

which is called the **Gell-Mann and Brueckner ground-state energy.** This formula corresponds to Eq. (1.11) in the classical plasma and is exact in the limit of small coupling constant $r_s$.

Wigner (1934, 1938) pioneered the study on the ground-state energy of a degenerate electron system. He treated correlations in high-density electron system by the variational method and adopted the ion-sphere model of Section 2.4 for the low-density electron system. He thereby interpolated the calculations carried out in both limits to obtain

$$E_c(r_s) = - \frac{0.88}{r_s + 7.8} \quad [\mathrm{Ry}] \tag{6.46}$$

as an expression for the correlation energy per electron. Isihara and Montroll (1971) later advanced an improved calculation of this interpolation formula.

The correlation energy can be calculated as a function of $r_s$ on the basis of various dielectric response functions mentioned in this chapter. In the strong-coupling domain this should make a significant improvement over the RPA calculation. We summarize results of such calculations in Fig. 6.5.

Recently Ceperley and Alder (1980) carried out Green's function Monte Carlo (GFMC) calculations for $E_c(r_s)$ at $r_s = 1$, 2, 5, 10, 20, 50, and 100. Vosko, Wilk, and Nusair (1980) then

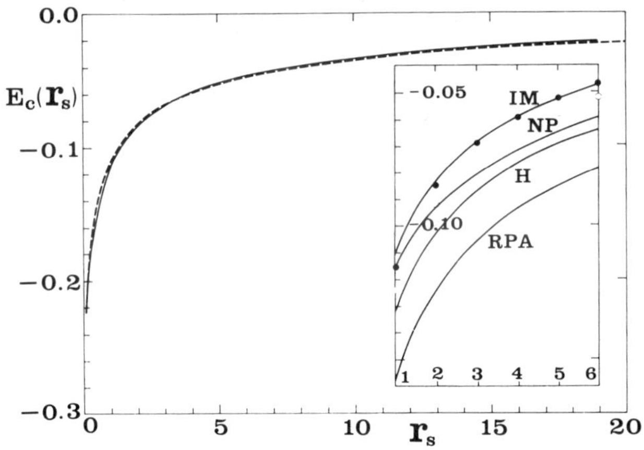


**Figure 6.5.** Correlation energy $E_c(r_s)$ vs. $r_s$. After Utsumi and Ichimaru (1980b). The solid curve represents the values obtained in Utsumi and Ichimaru (1980b), and the dashed curve refers to the Isihara-Montroll [4, 2] formula (1971). In the inset, the predictions of various theories are compared: RPA is the theory with $G(k) = 0$, i.e., Eq. (6.16); H with Eq. (6.33); NP in Nozières and Pines (1958); IM in Isihara and Montroll (1971); the open circles, the STLS result with Eq. (6.39); and the solid circles, the result of Utsumi and Ichimaru (1980b).

derived an interpolation formula for those data through a Padé approximant technique. The formula so fitted for the paramagnetic fluid phase reads

$$r_s \frac{dE_c(r_s)}{dr_s} = b_0 \frac{1 + b_1 x}{1 + b_1 x + b_2 x^2 + b_3 x^3} \tag{6.47}$$

where $x = \sqrt{r_s}$, $b_0 = 0.0621814$, $b_1 = 9.81379$, $b_2 = 2.82224$, and $b_3 = 0.736411$. Ichimaru and Utsumi (1981; Utsumi and Ichimaru 1982) obtained a parametrized expression for the local-field correction $G(\vec{k})$, which accurately reproduces the GFMC results and other microscopic calculations.

Those recent developments and comparisons between theories are described in e.g., Singwi and Tosi (1981) and Ichimaru (1982).

## Chapter 7

# Charged Particles or X-Rays Injected in Metals

In the previous chapter we studied the fundamental properties of the electron plasma in metal. Those properties can be experimentally investigated by means of scattering of incident X-rays or electron beams. The spectrum of electron-density fluctuations such as the plasma oscillations can be monitored through the energy-loss spectroscopy in a transmission experiment. It is also possible to probe the microscopic features of electron correlations through injection of molecular-ionic beams into metal. In this chapter we treat the properties of electron plasmas in metal which can be detected by those scattering methods.

## 7.1 Characteristic Energy-Loss Spectroscopy

In Section 3.4 we considered the differential cross section of the electromagnetic-wave scattering by plasma and showed that the spectral and angular distributions of scattered waves are directly connected with the spectrum of electron-density fluctuations in the plasma. In Chapter 4 we took up the radar backscattering experiments from the ionosphere and considered the theoretical implication of the results. Analogous scattering experiments have been performed extensively on the electron plasma in metal, as well, and have thereby unveiled valuable information on the excitation spectrum of density fluctuations. In this, and the following sections, we shall outline the method and results of those experiments.

The X-ray scattering spectroscopy is schematically illustrated in Fig. 7.1. The characteristic X-rays of a metal are in many cases used as the incident X-rays. For example, the $K\alpha_1$ line of copper has the energy 8.052 keV (1.540 Å in wavelength) and the natural width 2.5 eV; the $K\alpha_2$ line of copper has the energy 8.031 keV (1.544 Å in wavelength) and the natural width 3.5 eV. Transmitting such an X-ray into a thin metallic plate, one measures the spectral distribution of the scattered X-ray with an angle $\theta$ relative to the incident direction. The differential cross section of scattering is given by Eq. (3.15); the energy and momentum of the incident X-ray, $\hbar\omega_1$ and $\hbar\vec{k}_1$, and those of the scattered X-ray, $\hbar\omega_2$ and $\hbar\vec{k}_2$, are connected via Eqs. (3.7) and (3.8).

The energy difference $\hbar\omega$ between the incident and scattered

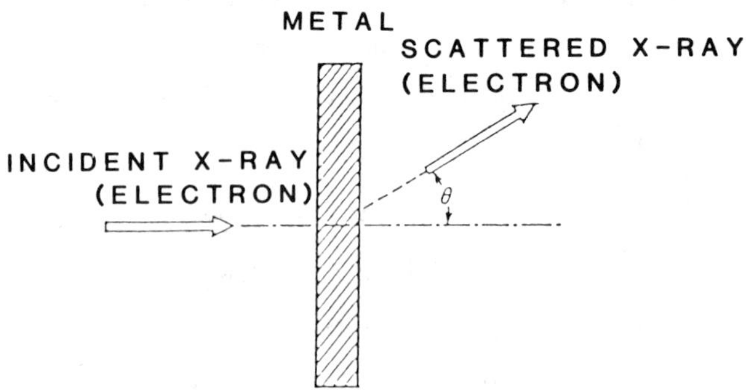


**Figure 7.1.** X-ray (electron) transmission-scattering spectroscopy.

X-rays corresponds to the excitation energy of a density fluctuation in the electron plasma. The plasma oscillation is a typical example of such collective density flucutations; its quantum, called a **plasmon**, has the energy

$$\hbar\omega_p = 47.1\ r_s^{-3/2}[\text{eV}] \ . \tag{7.1}$$

Although this expression does not take account of any effective-mass correction to the metallic electrons or the dielectric constant of the ionic lattice, we may approximately estimate $\hbar\omega_p = 3.2 - 16.7$ eV, for the electrons at metallic densities ($r_s = 2 - 6$). The numerical examples cited above indicate $\hbar\omega_1 \gg \hbar\omega_p$, so that we may take $\hbar\omega_1 \simeq \hbar\omega_2$. Hence for $\theta \ll 1$,

$$k = (\omega_1/c)\theta \ . \tag{7.2}$$

In these transmission-scattering experiments, one can alternatively use a beam of mono-energetic electrons (electron energy-loss spectroscopy) instead of the X-ray. In this case a metal foil with a thickness of around $10^3$ Å is used; its differential cross section of scattering takes a form analogous to Eq. (3.15), i.e.,

$$\frac{d^2Q}{d\Omega d\omega} = \frac{Vm^2}{4\pi^2\hbar^4}\left(\frac{E_1}{E_2}\right)^{1/2}\left(\frac{4\pi e^2}{k^2}\right)^2 S(\vec{k},\ \omega) \tag{7.3}$$

(e.g., Pines 1964; Ichimaru 1973). Denoting by $\vec{p}_1$, $E_1$, $\vec{p}_2$, and $E_2$ the momentum and energy of the incident and scattered electrons, we have relations corresponding to Eqs. (3.7) and (3.8) as

$$\hbar\omega = E_2 - E_1 \tag{7.4}$$

$$\hbar\vec{k} = \vec{p}_2 - \vec{p}_1 \ . \tag{7.5}$$

Those relations connect between the energy and momentum of the incident and scattered particles and the excitation energy $\hbar\omega$ and momentum $\hbar\vec{k}$ of the density fluctuations.

In an actual experiment, the energy range of 20 - 300 KeV is used for $E_1$; hence, as in the case of the X-ray scattering, one can take $E_1 \simeq E_2$. For $\theta << 1$, one thus has

$$k = \frac{1}{\hbar}\sqrt{2mE_1}\ \theta \ . \tag{7.6}$$

The energy resolution of the scattered particles, depending on the cases of the experiment, is usually 0.1 - 1 eV.

## 7.2 Plasmon Dispersion

Figure 7.2 shows an example of the energy-loss spectrum measured in an electron transmission experiment (Gibbons, Schnatterly, Ritsko, and Fields 1976). The sample is an aluminum foil with thickness 1100 Å. As we have noted earlier, aluminum has $r_s = 2.1$, the Fermi wave number $k_F = 1.73$ Å$^{-1}$ by Eq. (6.2), the RPA cutoff wave number $k_c = 1.2$ Å$^{-1}$ by Eq. (6.24), and the plasmon energy $\hbar\omega_p = 15.5$ eV by Eq. (7.1). All the data in Fig. 7.2 are confined in the wave number domain $k < k_c$; the spectral peak can be regarded as corresponding to the plasmon excitation. The peak energy value is observed to increase with the wave number; the dispersion relation of the plasma oscillation can be determined experimentally through such an observation. One can likewise estimate the decay rate of plasma oscillations from the half width of the peak structure.

We define the **plasmon dispersion coefficient** $\alpha$ by expressing the measured plasmon energy as a function of the scattering angle $\theta$:

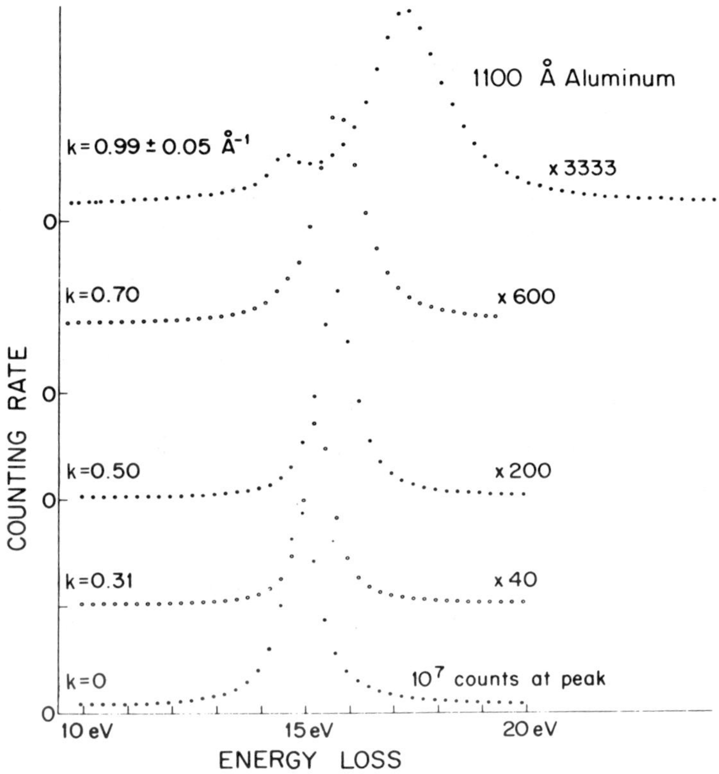


**Figure 7.2.** Electron energy-loss spectra of aluminum at various momentum transfers. After Gibbons et al. (1976).

$$\hbar\omega(\theta) = \hbar\omega_p + 2\alpha E_1\theta^2 . \tag{7.7}$$

Since the plasmon wave number and the scattering angle are related via Eq. (7.6), Eq. (7.7) implies

$$\hbar\omega_k = \hbar\omega_p + \alpha \frac{\hbar^2}{m} k^2 . \tag{7.8}$$

In the RPA, one then has

$$\alpha_{RPA} = \frac{3}{5}\frac{E_F}{\hbar\omega_p} = 0.637\ r_s^{-1/2} \tag{7.9}$$

in view of the dispersion relation Eq. (6.20).

The measured values of $\alpha$ for various metals are summarized in Fig. 7.3 on the basis of the data compiled by Raether (1980). The values are scattered rather widely partly because of different experimenters and experimental uncertainties. Nevertheless the observed results appear to indicate a consistent deviation from $\alpha_{RPA}$ based on the weak coupling theory; the difference obviously tends to widen as $r_s$ increases.

A number of theoretical attempts have been advanced to calculate the dispersion coefficient $\alpha$ by taking account of the strong coupling effects. For example, the plasmon dispersion

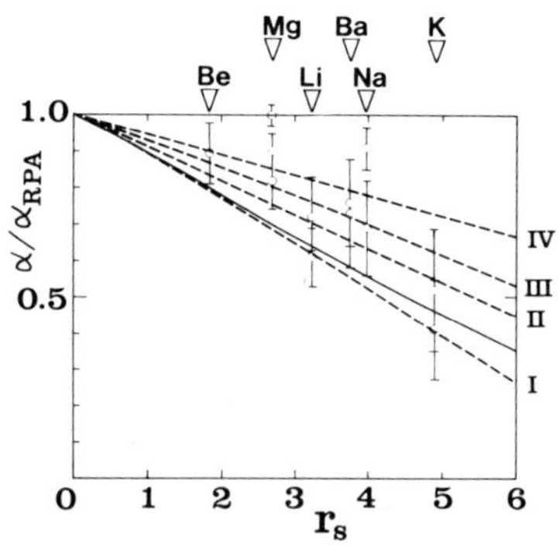


**Figure 7.3.** Plasmon dispersion coefficient $\alpha$ divided by RPA value $\alpha_{RPA}$ vs. $r_s$ in various approximations. Solid curve corresponds to the result of Utsumi and Ichimaru (1981). Dashed curves I-IV correspond to $\alpha_0$, $\alpha_{TW}$, $\alpha_\infty$, and $\alpha_{NP}$ in Eqs. (7.10), (7.11), (7.15), and (7.16), respectively. The experimental values of open circles with error bars are taken from Raether (1980). After Utsumi and Ichimaru (1981).

coefficient calculated on the basis of Eq. (6.30), which accounts for the static local-field correction effect, is

$$\alpha_0 = \alpha_{RPA} - \frac{\hbar\omega_p}{4E_F} \gamma_0(r_s) \tag{7.10}$$

where $\gamma_0(r_s)$ is defined in terms of the long-wavelength limit of $G(\vec{k})$ by Eq. (6.34). As we noted in Section 6.4, the calculational method for $G(\vec{k})$ varies depending on the approximations adopted; consequently the evaluation of $\gamma_0(r_s)$ also depends on the theory. In the Hartree-Fock limit of $r_s \to 0$, the Coulomb interaction may be neglected; Eq. (6.38) is applicable because only the exchange interaction remains. Equation (7.10) thus becomes

$$\alpha_{TW} = \alpha_{RPA} - \frac{\hbar\omega_p}{16E_F} . \tag{7.11}$$

This expression was derived by Toigo and Woodruff (1971).

The local-field correction $G(\vec{k})$ is introduced so that the low-frequency properties of the dielectric response function are accurately described. In the high-frequency domain, the strong coupling effect takes a different form. Writing the high-frequency limit of $\psi(\vec{k}, \omega)$ in Eq. (6.28) as

$$\psi(\vec{k}, \omega) \xrightarrow[(\omega \to \infty)]{} \frac{4\pi(Ze)^2}{k^2} [1 - G_\infty(\vec{k})] , \tag{7.12}$$

we find that the function $G_\infty(\vec{k})$ in the long-wavelength domain takes the form

$$\lim_{k\to 0} G_\infty(\vec{k}) = \gamma_\infty(r_s) \frac{k^2}{k_F{}^2} \tag{7.13}$$

where

$$\gamma_\infty(r_s) = \frac{1}{5}\int_0^\infty \frac{dk}{k_F}\,[1 - S(\vec{k})] \ . \tag{7.14}$$

The limiting form Eq. (7.12) is derived from the **third frequency-moment sum rule** (Puff 1965; Pathak and Vashishta 1973; Ichimaru 1982) of the dielectric function. Use of Eq. (7.12) in Eq. (6.28) yields the dispersion coefficient,

$$\alpha_\infty = \alpha_{RPA} - \frac{\hbar\omega_p}{4E_F}\,\gamma_\infty(r_s) \ . \tag{7.15}$$

In the Hartree-Fock limit, $r_s \to 0$, we may substitute Eq. (A6.2) in Eq. (7.14), to find

$$\alpha_{NP} = \alpha_{RPA} - \frac{3\hbar\omega_p}{80E_F} \ . \tag{7.16}$$

This expression was obtained by Nozières and Pines (1958) and by DuBois (1959).

Since the actual plasma frequency $\omega_p$ takes on an intermediate value, the value of $\alpha$ should be different from either of those limiting values. Theoretical values of $\alpha$ obtained through those various methods of approximation are exhibited in Fig. 7.3. The experimental values being scattered widely, it is not possible to judge which of those approximations is the most appropriate one. We may conclude, however, that the general trend of the experimental results are collectively described by the theories.

Let us revisit the data in Fig. 7.2, all of which correspond to the cases with $k < k_c$. We note that the decay rate of the plasmon determined from the half width of the peak does not vanish and increases with the wave number. As we saw in Section 6.3, the decay rate of plasmon in the RPA should vanish because no Landau damping is operative for $k < k_c$. Consequently we interpret at least a part of the decay rate observed in Fig. 7.2 as stemming from the strong coupling effects in the electron plasmas. One such strong coupling effects is the plasmon decay due to Coulomb collisions between electrons (cf. Section 2.8).

## 7.3 Stopping Power and the Wake Potential

In gaseous plasmas such as those produced by electric discharge, one can investigate the features of density correlations by inserting probes such as a grid, in addition to utilizing scattering of electromagnetic waves. We took up an example of that method in Section 3.1. For electron plasma in metal, however, a probe method is not useful because the correlation distance here is much shorter than those in gaseous plasmas. Energetic ions injected externally into metal are found to play, in effect the role of a probe, whereby the correlation properties in the electron plasma are experimentally investigated. In this, and, next sections, we shall consider these methods (see also e.g., Ohtsuki 1983).

For preparation we begin with a calculation of the force exerted onto an ion with an electric charge Ze and mass M traveling at a velocity $\vec{u}$ in the degenerate electron plasma. We assume that u is much greater than the Fermi velocity $v_F$ so that

$$u \gg v_F \ . \tag{7.17}$$

For a proton, Eq. (7.17) implies

$$\frac{1}{2} M u^2 \gg 92.1 \ r_s^{-2} \quad (\text{keV}) \ .$$

The injected charge produces a potential of external disturbance in the notation of Section 6.2. Choosing the position of the ion at $t = 0$ as the origin of the coordinates, we express the associated charge density as $Ze\delta(\vec{\mathbf{r}} - \vec{\mathbf{u}}t)$. Its space-time Fourier transformation then yields

$$Ze(\rho_{ext})_{\vec{\mathbf{k}}\omega} = Ze \int_{-\infty}^{\infty} dt \int_{\text{unit volume}} d\vec{\mathbf{r}}\delta(\vec{\mathbf{r}} - \vec{\mathbf{u}}t)\exp[-i(\vec{\mathbf{k}} \cdot \vec{\mathbf{r}} - \omega t)]$$

$$= 2\pi Ze\delta(\omega - \vec{\mathbf{k}} \cdot \vec{\mathbf{u}}) \ . \tag{7.18}$$

To calculate $(\phi_{ext})_{\vec{\mathbf{k}}\omega}$ in Eq. (6.8), we multiply Eq. (7.18) by $4\pi/k^2$, in accord with the Poisson equation:

$$(\phi_{ext})_{\vec{\mathbf{k}}\omega} = 8\pi^2 \frac{Ze}{k^2} \delta(\omega - \vec{\mathbf{k}} \cdot \vec{\mathbf{u}}) \ . \tag{7.19}$$

The injected ion induces the electron-density fluctuations $\langle\rho_{ind}\rangle_{\vec{\mathbf{k}}\omega}$ in the plasma. This quantity can be calculated by setting $Z = -1$ in Eq. (6.10). The space-time distribution $\rho_{ind}(\vec{\mathbf{r}}, t)$ of the induced electron-density fluctuations is obtained by carrying out the inverse Fourier transformation of $\langle\rho_{ind}\rangle_{\vec{\mathbf{k}}\omega}$; the result is

$$\rho_{ind}(\vec{\mathbf{r}}, t) = Z \int \frac{d\vec{\mathbf{k}}}{(2\pi)^3} \left[1 - \frac{1}{\varepsilon(\vec{\mathbf{k}}, \vec{\mathbf{k}}\cdot\vec{\mathbf{u}})}\right]\exp[i\vec{\mathbf{k}} \cdot (\vec{\mathbf{r}} - \vec{\mathbf{u}}t)] \ . \tag{7.20}$$

The condition (7.17) leads to the expansion, Eq. (6.19), of the dielectric response function, of which we retain the first two terms on the right-hand side. The function $\varepsilon(\vec{k}, \omega)$ has a small imaginary part, which is in general an odd function of $\omega$ as Eq. (6.21) illustrates. Consequently we may adopt an approximation*

$$\varepsilon(\vec{k}, \omega) = 1 - \frac{\omega_p^2}{\omega^2} + i\eta \frac{\omega}{|\omega|} \tag{7.21}$$

for the dielectric function to be used in Eq. (7.20), where $\eta$ is a positive infinitesimal.

We calculate Eq. (7.20) by choosing the x axis in the $\vec{u}$ direction. The integrations with respect to $k_x$ and $k_y$ in Eq. (7.20) then produce $\delta(x)$ and $\delta(y)$. To carry out the $k_z$ integration, we close the contour of integration by an infinite semicircle with $\mathrm{Im}k_z > 0$ on the complex $k_z$ plane when $z - ut > 0$ (see Fig. 7.4). The integral pertaining to Eq. (7.20)

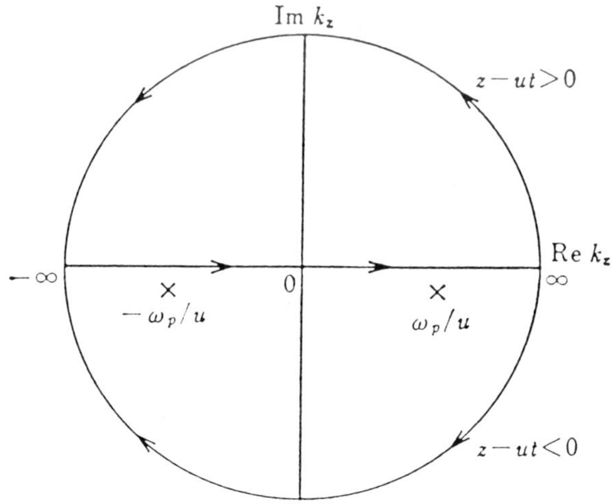


**Figure 7.4.** The contours of $k_z$ integration in Eq. (7.20).

---

* This expression is applicable irrespective of the magnitude of the Coulomb coupling constant $r_s$.

vanishes by virtue of Cauchy's theorem, because no roots of $\varepsilon(\vec{\mathbf{k}}, k_z u) = 0$ exist on the upper half of the complex plane for Eq. (7.21). For $z - ut < 0$, on the other hand, we close the contour of integration by an infinite semicircle with $\mathrm{Im} k_z < 0$. In this case the residues at the two roots of $\varepsilon(\vec{\mathbf{k}}, k_z u) = 0$ in the lower half plane contribute to the integration as shown in Fig. 7.4. As a result we find

$$\rho_{ind}(\vec{r}, t) = -\frac{Z}{\lambda_p}\, \delta(x)\delta(y)\sin\left(\frac{z - ut}{\lambda_p}\right)\theta(ut - z) \, . \qquad (7.22)$$

Here $\theta(x)$ is the unit step function

$$\theta(x) = \begin{cases} 1 & (x \geq 0) \\ 0 & (x < 0) \end{cases} \qquad (7.23)$$

and

$$\lambda_p = u/\omega_p \qquad (7.24)$$

represents the characteristic length corresponding to the distance ($= 2\pi\lambda_p$) over which an ion travels during one period of the plasma oscillation.

Equation (7.22) contains a number of interesting features. First we remark on the involvement of the step function $\theta(ut - z)$: Since $ut$ is the $z$ coordinate of the injected ion, the electron-density induced by this ion is limited to the domain $z < ut$ behind it; the effect does not reach the domain $z > ut$ where the ion will travel subsequently. This is a manifestation of the principle of causality mentioned in connection with Eqs. (6.8) and (6.9).

The induced electron density thus varies with wavelength and frequency corresponding to the plasma oscillation in the domain posterior to the injected ion along the z axis. These are the features of correlations characterizing the interaction between the injected ion and the degenerate electrons. We note a remarkable difference between these and the static screening treated in Section 2.2.

We next take up the problem associated with $\delta(x)\delta(y)$ contained in Eq. (7.22). This function appears because we regarded the injected ion as a point charge and calculated the electron-density variations induced by collisions with the ion. Because of the condition (7.17), the spatial diffusion of electrons need not be considered; the variation of the electron density is confined on the z axis. In the treatment of close encounters between the ion and the electrons, one must regard the particles, not as classical point charges, but as those having quantum-mechanical spread. In the center of the mass system of scattering between the ion and an electron, we may take the reduced mass to be equal to the electron mass; the relative velocity is u because of the condition (7.17). Hence the relevant momentum is mu with the corresponding de Broglie wavelength of $\hbar/mu$. The induced electron density should have a spread of this magnitude around the z axis.

The space-charge density induced in the plasma produces an electrostatic potential associated with it. Taking account of the aforementioned effect of quantum-mechanical spread in Eq. (7.22), Vager and Gemmel (1976) calculated the potential as

$$\phi_{ind}(R, z - ut) = -\frac{Ze}{\lambda_p}\int_0^\infty d\zeta \frac{\sin(\zeta/\lambda_p)}{\sqrt{R^2 + (\hbar/mu)^2 + (\zeta + z - ut)^2}} \tag{7.25}$$

where $R = \sqrt{x^2 + y^2}$ is the distance from the z axis. Since Eq. (7.25) represents the electrostatic potential produced by excitation of plasma waves as an ion travels through the plasma, we shall call it the **wake potential.** Figure 7.5 illustrates such a wake potential computed from Eq. (7.25) when a 400 keV proton is injected in a carbon plasma ($\hbar\omega_p$ = 25.0 eV).

As we observe in the gradient of the potential at the position of the ion in Fig. 7.5, the injected charged particle suffers a retarding force from the wake potential. The **stopping power** of the plasma is then the rate, -dw/dz, at which the injected charged particle loses its kinetic energy $w( = \frac{1}{2} Mu^2)$ per unit length owing to this retardation effect. This rate is thus calculated from Eq. (7.25) as

$$-\frac{dw}{dz} = Ze \left.\frac{\partial \phi_{ind}}{\partial z}\right|_{\substack{z=ut \\ R=0}} \simeq \frac{Z^2 e^2 \omega_p^2}{u^2} \ln \left(\frac{1.123\ mu^2}{\hbar\omega_p}\right) . \qquad (7.26)$$

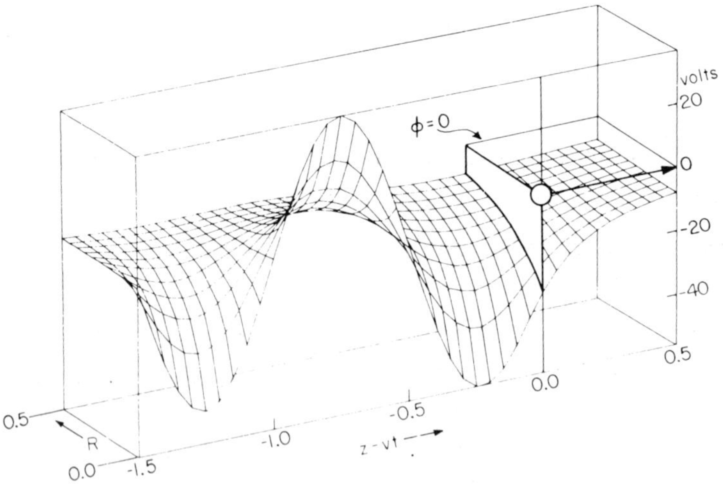


**Figure 7.5.** Potential distribution Eq. (7.25) associated with the potential wake of a 400-keV proton traversing carbon ($\hbar\omega_p$ = 25.0 eV). Distances are shown in units of $2\pi\lambda_p$ = 14.5 Å. After Vager and Gemmell (1976).

Here $\lambda_p >> \hbar/mu$ is assumed, and $2\exp(-\gamma) = 1.123$ where $\gamma$ is Euler's constant. Equation (7.26) is the **Bethe formula** for the stopping power; the logarithmic factor contained in it is another example of the Coulomb logarithm discussed in Section 2.3.

## 7.4 Ion Clusters Injected in Metals

Let us consider an experiment in which a beam of molecular ions such as $H_2^+$ with an energy of ~ 100 keV per nucleon is injected into a metal foil with a thickness of around 1000 Å (Fig. 7.6). The electrons which bind the two protons in the $H_2^+$ molecules will be removed soon after the molecules enter the metal. The dissociated two protons repel each other by the mutual electrostatic forces, depart from each other while traveling through the metal foil, and leave from the opposite side of the foil. Through such an experiment one can neatly monitor the dynamic correlation effects between the ion and the electron system in

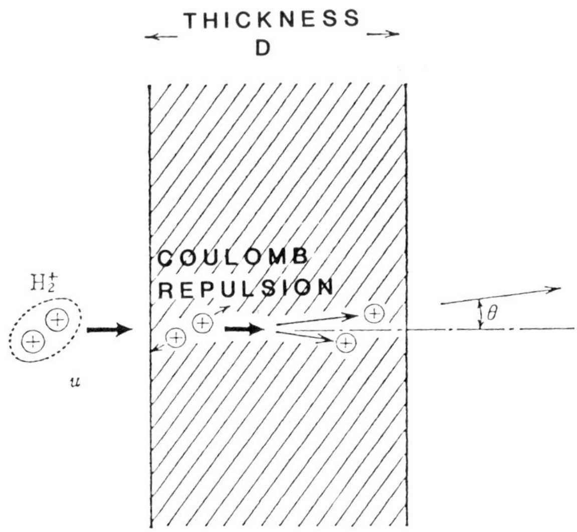


**Figure 7.6.** Ion cluster injected in metal.

metal as described in the previous section. In this section we look at the salient features of such an experiment.

We first take up the stopping-power experiment carried out by Brandt, Ratkowski, and Ritchie (1974). The principal issue of the experiment is the following: Recalling Eq. (7.26) we note that the stopping power of a plasma against an ion is proportional to the square of Z, the ionic charge. Then, does a molecular ion with electrons stripped off receive a stopping power as a single charge of $Z = 2$, or does it receive a stopping power as two individual charges with $Z = 1$ each? If the former is the case, the stopping power per nucleon will be twice that of the latter case. Consequently, a measurement of the energy loss in the charged particles leaving the foil will determine such an effective electric charge experimentally.

Figure 7.7 shows the effective electric charge $Z_{eff}$ measured by the stopping power when a molecular beam of $H_2^+$ or $H_3^+$ is injected into a gold foil or a graphite (carbon) foil with various thickness D. $\tau = D/u$ is the "dwell" time of an ion in the foil, and $\bar{\tau}$ refers to an average time for the protons to separate from each other at a distance of $\lambda_p = u/\omega_p$ by the electrostatic repulsion. When $\tau$ becomes considerably greater than $\bar{\tau}$, $Z_{eff}$ approaches unity. When $\tau$ is small, $Z_{eff} > 1$; $Z_{eff}$ for $H_3^+$ is greater than that for $H_2^+$.

For interpretation of the experimental results in Fig. 7.7, we must investigate how the ionic charge influences the stopping power through the scattering of electrons. As may be clear from the calculation leading to the integral form of Eq. (7.25), the electric field perceived by the ion is a superposition of fields produced by electron distribution at various distances. We classify those distances into two groups: short-range scattering around $\hbar/mu$ and long-range encounters around $u/\omega_p$. Note that the Coulomb logarithm in Eq. (7.26) has

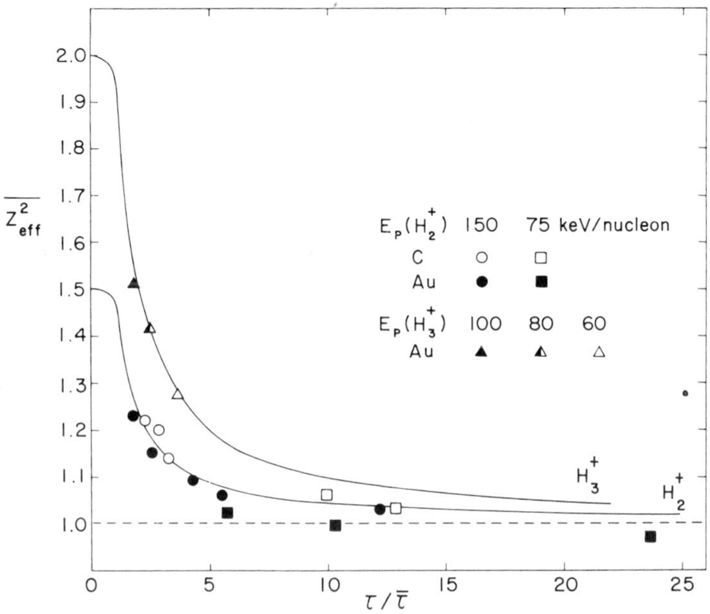


**Figure 7.7.** Measured energy-loss ratios $\Delta E(H_2^+)/2\Delta E(H^+)$ and $\Delta E(H_3^+)/3\Delta E(H^+)$ in various foils as a function of the dwell time $\tau$ in the foil, in units of $\bar{\tau}$. After Brandt, Ratkowski, and Ritchie (1974).

the ratio between those two characteristic distances as its argument. Whether a given ion cluster behaves as a single particle or as separate individual particles depends on the spread of the ion cluster relative to the distances of relevant electrons.

As for the spread of an ion cluster, let us remark that the average distance between two protons in a molecular ion $H_2^+$ is already 1.29 Å. We compute, on the other hand,

$$\frac{\hbar}{mu} = 0.265 \left[\frac{W}{100 \text{ keV}}\right]^{-1/2} \quad (\text{Å}) \tag{7.27}$$

where w is the energy per nucleon of the incident ion. Since this distance is smaller than the inter-proton distances in $H_2^+$ and $H_3^+$, those molecular ions behave as separate individual protons from the moment when the molecular ion is injected into the metal and its orbital electrons are removed, as far as the scattering with short-range electrons is concerned. The inter-proton distance being substantially smaller than $u/\omega_p$ as yet, however, the cluster of protons acts as if it were a single charged particle as far as those long-range scattering acts are concerned.

The solid curves in Fig. 7.7 depict the results of theoretical computation where those effects are taken into consideration. In the limit of $\tau/\bar{\tau} \to 0$, $Z_{eff}^2$ of $H_2^+$ and $H_3^+$ take on values, 1.5 and 2.0, respectively, with the reasons explained above. Take $H_2^+$ for example. From the long-range scattering contribution, $Z_{eff}^2 = (2)^2/2 = 2$ because the ion cluster behaves as a charged particle with $Z = 2$. From the short-range scattering contribution, $Z_{eff}^2 = (1^2 + 1^2)/2 = 1$ because the cluster behaves as two charged particles with $Z = 1$ each. Consequently the average between those two computations gives $Z_{eff}^2 = 1.5$; similarly one finds $Z_{eff}^2 = 2.0$ in the case of $H_3^+$.

As the proton cluster travels through the metal, the inter-proton distance increases due to the Coulomb repulsion and the cluster begins to behave as separate individual protons for the long-range scattering as well. Hence $Z_{eff}^2$ approaches unity as $\tau/\bar{\tau}$ increases.

Let us again look at Fig. 7.6. While two protons produced by dissociation of a molecular ion $H_2^+$ travel in metal, each proton produces a wake potential of Fig. 7.5 in addition to its own Coulomb field. The wake potential exerts a large influence on the ion motion behind it; the front-running proton in Fig.

7.6 thus has a large effect on the motion of the trailing proton by the wake potential.

Figure 7.8 shows experimental results where such a wake-potential effect is clearly detected (Gemmell, et al., 1975). Here $(^4HeH)^+$ molecular ions with energy 4 MeV are injected into a gold foil with thickness 535 Å and the angular distribution of the transmitting protons is observed. The molecular ion is split into a proton and an α particle in the gold foil. If the ordinary Coulomb repulsion is operative between those two

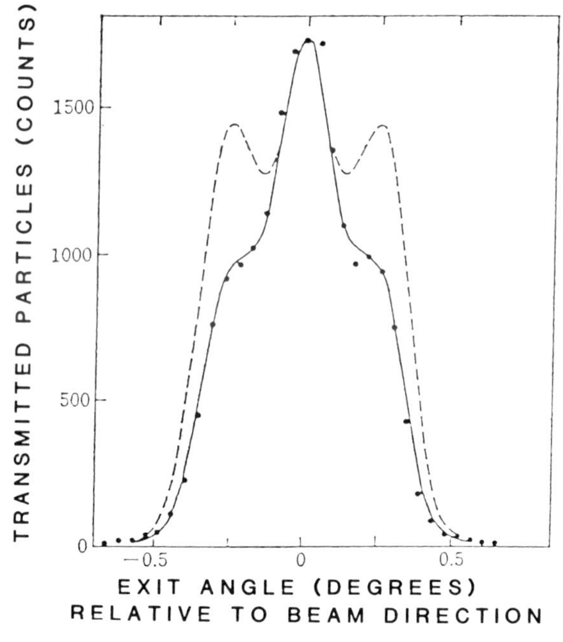


**Figure 7.8.** Angular distributions for protons from 4-MeV $(^4HeH)^+$, measured under (111)-planar-channeling conditions with a 535-Å-thick Au crystal. The points are the experimental data. The curves are calculated with (solid line) and without (dashed line) the wake potential. After Gemmell et al. (1975).

particles, then the angular distribution of protons theoretically expected would be that as shown by the dashed curve. The observed results, however, are as plotted by filled circles; they are obviously different from the dashed curve.

When the existence of the wake potential is taken into account, the front-running ion produces a force which acts to align the trailing ion along the z axis by reducing R. Consequently the aligning action is expected to work in such a way to shave off the two shoulders of the dashed curve in Fig. 7.8. A theoretical calculation carried out by taking an additional account of the wake-potential effect shown in Fig. 7.5 indeed produces the solid curve in Fig. 7.8, which correctly explains the experimental results. The microscopic correlation effects in the metallic electron plasma can thus be observed through such an experimental method.

# Chapter 8

# Electromagnetic Waves in Magnetized Plasma

We proceed to consider propagation of electromagnetic waves in a plasma with a uniform stationary magnetic field. The system is no longer isotropic, so that all the elements of the dielectric tensor (cf. Appendix 4) participate in the description of the propagation characteristics. Particularly notable is the emergence of low-frequency propagation modes along the magnetic field with phase velocities much less than the light velocity in a vacuum. In this chapter we look at some of the relevant experimental and theoretical results.

## 8.1 Helicon and Alfvén-Wave Experiments in Solid-State Plasmas

In 1961, Bowers, Legendy, and Rose (1961) reported the observation at 4 K of a novel galvanomagnetic effect in sodium of very high conductivity. In this effect, the Lorentz force across the magnetic field and the motion of the conduction electrons produced an electric oscillation of low frequency ranging from 5 to 35 Hz. These results provided a direct evidence of a particular electromagnetic excitation in a magnetized plasma, at a frequency much lower than both the plasma frequency and the cyclotron frequency, which was first predicted in 1960 by Aigrain (1961) and by Konstantinov and Perel' (1960); the word **helicon** was coined by Aigrain to describe such a propagation mode.

The experimental arrangement is shown schematically in Fig. 8.1. A cylindrical specimen of sodium (height 30 mm and diameter 4 mm) is placed inside primary and secondary coils. The ratio between the specimen electric resistivities at 300 K and 4 K is about 7500; the product $|\Omega_e|\tau$ between the electron cyclotron frequency* and the collisional mean free time is about 40 in $10^4$ gauss at 4 K. On closing or opening the primary circuit, eddy currents are induced in the specimen which can be detected by the secondary coil.

Figure 8.2 shows oscilloscope traces of the secondary voltage as a function of time on opening primary circuit for various values of the applied magnetic field B. In zero magnetic field the normal eddy-current decay is observed. The other three traces show the effect of applying B. One observes

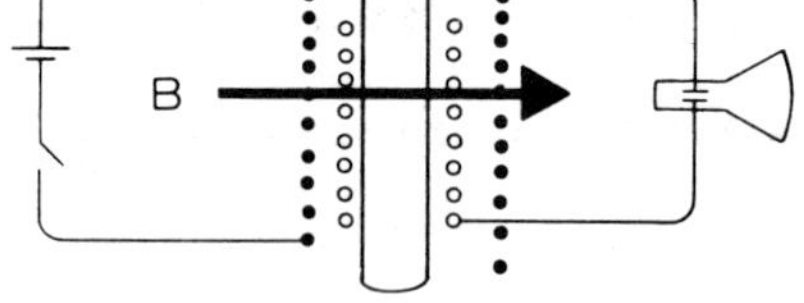


**Figure 8.1.** Schematic diagram of helicon-propagation experiment. After Bowers, Legendy, and Rose (1961).

---

* In this chapter we follow the notation of Eq. (A4.13) rather than Eq. (4.11).

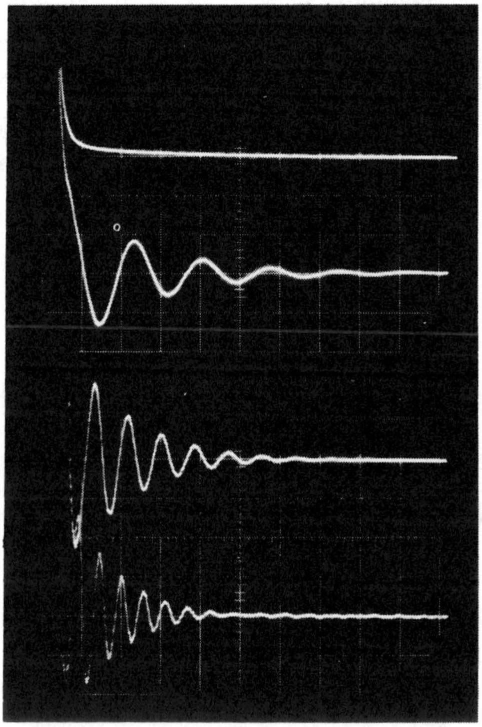

**Figure 8.2.** Voltage in secondary coil as a function of time on opening primary circuit for different values of B. Starting from the top, the four curves correspond to B = 0, 3600, 7200, and 10800 gauss, respectively. Ordinate is 5 millivolts per large division; abscissa is 50 milliseconds per large division. After Bowers, Legendy, and Rose (1961).

damped oscillations whose frequency is proportional to the magnetic field; the frequency at $10^4$ gauss is 32 Hz. These relations are consistent with the dispersion relation of the helicon, which we shall derive later in Section 8.4.

We now turn to another experiment demonstrating **Alfvén wave** propagation in bismuth (Williams and Smith 1964). Alfvén waves are magnetohydrodynamic waves which propagate in a magnetized plasma of high conductivity consisting of equal numbers of mobile positive and negative carriers. Their existence and properties were first predicted by Alfvén (1942). The Alfvén waves have been investigated in various substances, including gaseous-discharge plasmas (Allen, Baker, Pyle, and Wilcox 1959; Jephcott 1959; Wilcox, Boley, and DeSilva 1960), liquid metals (Lundquist 1949; Lehnert 1954), and solids (Buchsbaum and Galt 1961; Kaner and Skobov 1966; Glicksman 1971). The experiment by Williams and Smith (1964) has provided a good fit between theory and experiment for Alfvén wave propagation in a carrier-compensated semimetal.

In the experiment a microwave radiation is incident on the sample surface of bismuth with thickness, $d = 3.06$ mm. A microwave frequency ($\omega/2\pi$) was chosen in the range 12 - 18 GHz, in order to have $\omega\tau > 1$ but still allow $|\Omega_e| > \omega$ to be reached at reasonable magnetic fields. The "leakage" fringes, i.e., interference between the microwaves leaking around the sample and those which have passed once through the sample, are detected by the use of bolometric technique. Figure 8.3 exhibits such interference fringes at 13.54 GHz with the magnetic field normal to the surface and parallel to a bisectrix axis.

The optical thickness of the sample is calculated directly from the refractive index $\eta$, which is equal to the square root of the dielectric constant [cf. Eq. (4.9)]. The interference relationship is $N\lambda_0 = 2d\eta$, where $\lambda_0$ is the free space wavelength and N the fringe index, or number of half wavelengths in a sample thickness. The fringe index obtained for the data of Fig. 8.3 is plotted in Fig. 8.4 as a function of $1/B$. We here observe clearly that the fringe index N follows a straight line; it has also been found that the effective refractive index has no frequency dependence. These results are consistent with the

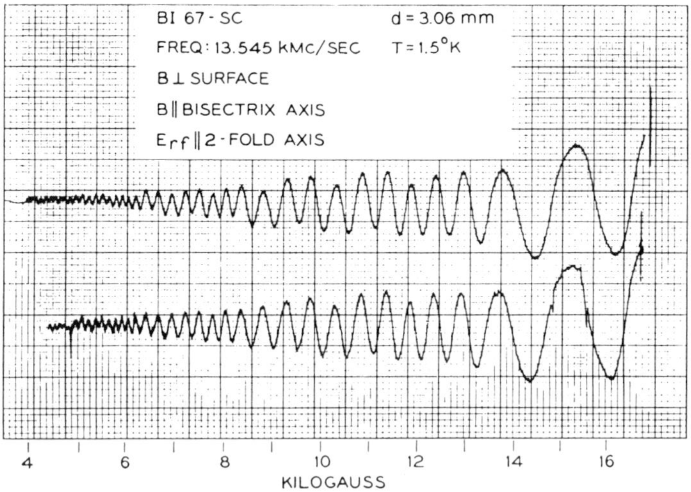


**Figure 8.3.** Interference fringes at 13.54 kMc/s in a sample of bismuth with the magnetic field normal to the surface and parallel to a bisectrix axis. After Williams and Smith (1964).

propagation of the Alfvén waves, which we shall take up later in Section 8.5.

Figure 8.4 also exhibits an oscillatory deviations of the data points from a straight-line behavior. These are the effects arising from the quantum oscillations of the Fermi surface, which are outside the scope of consideration in this volume.

## 8.2 Propagation of Electromagnetic Waves in a Magnetized Plasma

The dispersion relation for the propagation of the electromagnetic wave in a magnetized plasma may be analyzed generally through examination of the determinant equation (A8.6) in Appendix 8:

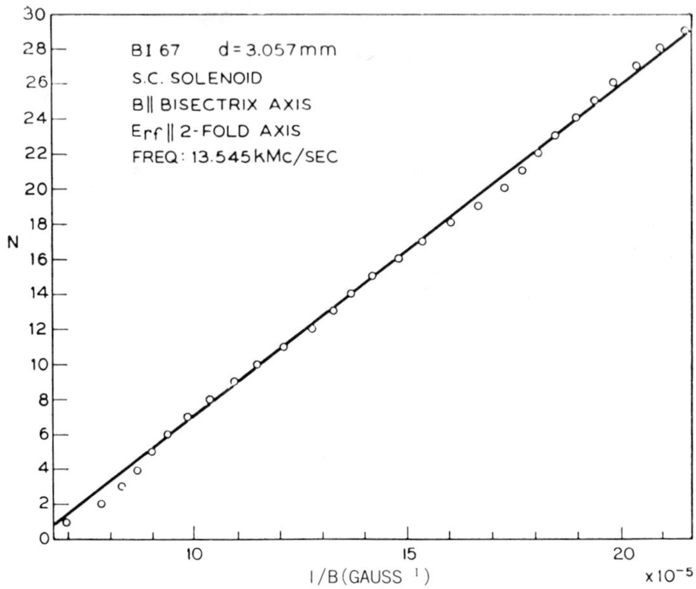


**Figure 8.4.** Plot of fringe index as a function of 1/B for the data of Fig. 8.3. The solid line is a least squares line through the data points. After Williams and Smith (1964).

$$\det\left|\overleftrightarrow{\varepsilon}(\vec{\mathbf{k}},\ \omega) - \left(\frac{kc}{\omega}\right)^2\left(\overleftrightarrow{I} - \frac{\vec{\mathbf{k}}\vec{\mathbf{k}}}{k^2}\right)\right| = 0 \ , \tag{8.1}$$

where the tensor $\overleftrightarrow{I}$ is defined by Eq. (A4.6). We assume the average macroscopic magnetic field in the z direction, i.e., $\langle\vec{\mathbf{B}}\rangle = B\hat{\mathbf{z}}$, and the particle distribution to be nonrelativistic Maxwellian, Eq. (1.2), at temperature T for each species $\sigma$. The dielectric tensor (A4.11) now takes a form,

$$\overleftrightarrow{\varepsilon}(\vec{\mathbf{k}}, \omega) = I - \sum_{\sigma} \frac{\omega_{p\sigma}^2}{\omega^2} \left\{ \sum_{\nu=-\infty}^{\infty} \frac{Z_0}{Z_\nu} \overleftrightarrow{P}_\sigma(\beta, Z_\nu;\nu)[1 - W(Z_\nu)] - Z_0^2 \hat{\mathbf{z}}\hat{\mathbf{z}} \right\} \tag{8.2}$$

where

$$\omega_{p\sigma}^2 = 4\pi n_\sigma e_\sigma^2/m_\sigma \tag{8.3}$$

$$Z_\nu = \frac{\omega - \nu\Omega_\sigma}{|k_\parallel|(k_B T/m_\sigma)^{1/2}} \tag{8.4}$$

and $W(Z_\nu)$ is the W function defined in Appendix 5. The tensor $\overleftrightarrow{P}_\sigma(\beta, Z_\nu;\nu)$ in Eq. (8.2) is calculated as

$$\overleftrightarrow{P}_\sigma(\beta, Z_\nu;\nu) = \begin{pmatrix} \frac{\nu^2}{\beta} \Lambda_\nu & i\nu\Lambda'_\nu & \frac{k_\parallel}{|k_\parallel|} \frac{\nu}{\sqrt{\beta}} Z_\nu \Lambda_\nu \\ - i\nu\Lambda'_\nu & \frac{\nu^2}{\beta} \Lambda_\nu - 2\beta\Lambda'_\nu & - i \frac{k_\parallel}{|k_\parallel|} \sqrt{\beta}\, Z_\nu \Lambda'_\nu \\ \frac{k_\parallel}{|k_\parallel|} \frac{\nu}{\sqrt{\beta}} Z_\nu \Lambda_\nu & i \frac{k_\parallel}{|k_\parallel|} \sqrt{\beta}\, Z_\nu \Lambda'_\nu & Z_\nu^2 \Lambda_\nu \end{pmatrix} \tag{8.5}$$

where $\beta = k_\perp^2 k_B T/m_\sigma \Omega_\sigma^2$, $\Lambda_\nu = I_\nu(\beta)\exp(-\beta)$, $\Lambda'_\nu = d\Lambda_\nu/d\beta$, and $I_\nu$ is the modified Bessel function of the $\nu$th order. As in Eq. (A4.12), the direction of the wave vector is set as $\vec{\mathbf{k}} = k_\perp \hat{\mathbf{x}} + k_\parallel \hat{\mathbf{z}}$.

## 8.3 Circularly Polarized Waves

Let us consider a special case of wave propagation along the magnetic field or in the z direction. In this case, $k_\perp = 0$

and $|k_\parallel| = k$, so that the tensor (8.5) takes a simplifed form,

$$\overleftrightarrow{P}_\sigma(0, Z_\nu;\nu) = \begin{bmatrix} \frac{1}{2}(\delta_{\nu,1} + \delta_{\nu,-1}) & \frac{i}{2}(\delta_{\nu,1} - \delta_{\nu,-1}) & 0 \\ -\frac{i}{2}(\delta_{\nu,1} - \delta_{\nu,-1}) & \frac{1}{2}(\delta_{\nu,1} + \delta_{\nu,-1}) & 0 \\ 0 & 0 & Z_0{}^2\delta_{\nu,0} \end{bmatrix} \tag{8.6}$$

where $\delta_{\nu,\nu'}$ refers to Kronecker's delta.

When an electromagnetic wave propagates in a plasma along the magnetic field, it turns out that a circularly polarized mode, rather than a linearly polarized mode, is a proper solution to the dispersion equation. It is thus useful to introduce a scheme of transformation from linearly polarized waves to circularly polarized waves.

The transformation may be achieved with the aid of a unitary matrix

$$\overleftrightarrow{U} = \begin{bmatrix} 1/\sqrt{2} & -i/\sqrt{2} & 0 \\ -i/\sqrt{2} & 1/\sqrt{2} & 0 \\ 0 & 0 & 1 \end{bmatrix}. \tag{8.7}$$

It satisfies the condition for unitarity

$$\overleftrightarrow{U} \cdot \overleftrightarrow{U}^{+} = \overleftrightarrow{U}^{+} \cdot \overleftrightarrow{U} = \overleftrightarrow{I}$$

where $\overleftrightarrow{U}^{+}$ is the Hermitian conjugate matrix to $\overleftrightarrow{U}$. An electric field vector expressed in terms of its Cartesian components is transformed as

$$\begin{bmatrix} E_r \\ E_\ell \\ E_z \end{bmatrix} = \overleftrightarrow{U} \cdot \begin{bmatrix} E_x \\ E_y \\ E_z \end{bmatrix}$$

or

$$E_r = (E_x - iE_y)/\sqrt{2} \tag{8.8a}$$

$$E_\ell = -i(E_x + iE_y)/\sqrt{2} \ . \tag{8.8b}$$

The new components $E_r$ and $E_\ell$ given by Eqs. (8.8) represent the right and left circularly polarized components, respectively.*

A tensorial quantity, such as the dielectric tensor (8.2), is transformed as $\overleftrightarrow{U} \cdot \overleftrightarrow{\varepsilon} \cdot \overleftrightarrow{U}^+$. The dielectric tensor after such a transformation is now expressed in a diagonal form

$$\overleftrightarrow{U} \cdot \overleftrightarrow{\varepsilon} \cdot \overleftrightarrow{U}^+ = \begin{bmatrix} \varepsilon_r & 0 & 0 \\ 0 & \varepsilon_\ell & 0 \\ 0 & 0 & \varepsilon_z \end{bmatrix} \ .$$

The diagonal elements for a multicomponent plasma may then be calculated, with the aid of Eq. (8.6), as

---

* This statement is true only when $\omega > 0$; we shall confine ourselves to this frequency domain throughout the remainder of this chapter.

$$\varepsilon_r(k, \omega) = 1 - \sum_\sigma \frac{\omega_{p\sigma}^2}{\omega(\omega + \Omega_\sigma)} [1 - W(\frac{\omega + \Omega_\sigma}{k(k_BT/m_\sigma)^{1/2}})] \tag{8.9a}$$

$$\varepsilon_\ell(k, \omega) = 1 - \sum_\sigma \frac{\omega_{p\sigma}^2}{\omega(\omega - \Omega_\sigma)} [1 - W(\frac{\omega - \Omega_\sigma}{k(k_BT/m_\sigma)^{1/2}})] \tag{8.9b}$$

$$\varepsilon_z(k, \omega) = 1 + \sum_\sigma \frac{k_{D\sigma}^2}{k^2} W(\frac{\omega}{k(k_BT/m_\sigma)^{1/2}}) \tag{8.9c}$$

where

$$k_{D\sigma}^2 = 4\pi n_\sigma e_\sigma^2/k_BT . \tag{8.10}$$

The dispersion relation (8.1) is now factored into a product of three equations, so that

$$\varepsilon_r(k, \omega) = (kc/\omega)^2 \tag{8.11a}$$

$$\varepsilon_\ell(k, \omega) = (kc/\omega)^2 \tag{8.11b}$$

$$\varepsilon_z(k, \omega) = 0 . \tag{8.11c}$$

The first two equations are the dispersion relations for the right and left circularly polarized waves, which we shall examine for specific cases in the subsequent sections. Equation (8.11c) is the dispersion relation for the longitudinal mode, equivalent to Eq. (6.15).

## 8.4 Uncompensated Plasmas

We first take up a special case of a plasma in which the densities of oppositely charged mobile carriers may not exactly compensate each other. Such a situation occurs frequently in solids; a typical example is the conduction electrons in metal. In this section we assume that the uncompensated plasma is a one-component plasma consisting of electrons.

The dielectric function (8.9a) for the right circularly polarized wave in these circumstances reads

$$\varepsilon_r(k, \omega) = 1 - \frac{\omega_{pe}^2}{\omega(\omega - |\Omega_e|)} \left[1 - W\left(\frac{\omega - |\Omega_e|}{k(k_B T/m_e)^{1/2}}\right)\right] . \qquad (8.12)$$

In the light of Eq. (A5.25), the imaginary part of this dielectric function will be negligible if

$$||\Omega_e| - \omega| \gg k(k_B T/m_e)^{1/2} . \qquad (8.13)$$

With this assumption the dispersion relation (8.11a) becomes

$$\left(\frac{kc}{\omega}\right)^2 = 1 - \frac{\omega_{pe}^2}{\omega(\omega - |\Omega_e|)} . \qquad (8.14)$$

This relation is illustrated by the solid curves in Fig. 8.5.

The cutoff and resonance frequencies, discussed in Section 4.1, may be calculated from Eq. (8.14). We find the cutoff frequency at

$$\omega_{cr} = \frac{1}{2} \left[(\Omega_e^2 + 4\omega_{pe}^2)^{1/2} + |\Omega_e|\right] . \qquad (8.15)$$

The resonance would take place at the frequency

$$\omega_R = |\Omega_e| . \qquad (8.16)$$

In the magnetic field, the electrons gyrate in the same direction as the right circularly polarized wave; a resonance is expected when the wave frequency matches the cyclotron frequency.

When the thermal velocities of the electrons are finite, however, the condition (8.13) may be violated in the vicinity of the resonance frequency Eq. (8.16). Thus, we expect that in the domain

$$|\Omega_e| - k(k_B T/m_e)^{1/2} < \omega < |\Omega_e| + k(k_B T/m_e)^{1/2} \tag{8.17}$$

the imaginary part of Eq. (8.12) and consequently the damping of the wave will be substantial. The physical origin of this attenuation can be easily understood: When a wave with frequency $\omega$ and wave number k propagates along the magnetic field in the plasma, an electron with a parallel velocity $v_\parallel$ experiences, because of the Doppler effect (see Section 3.3), a shifted frequency

$$\omega' = \omega \pm kv_\parallel \ .$$

If $\omega'$ coincides with the electron cyclotron frequency, then the resonant absorption of the wave energy by the electron will take place; Eq. (8.17) represents the condition for availability of many such electrons. This phenomenon, called the **Doppler-shifted cyclotron resonance,** is therefore related closely to the Landau damping of the plasma oscillations considered in Section 2.7.

A most remarkable feature in the dispersion relation (8.14) is the emergence of a new propagation mode at low frequencies. In the low-frequency domain, we may neglect the first term, unity, on the right-hand side of Eq. (8.14), so that

$$\omega = \frac{|\Omega_e|(kc)^2}{\omega_{pe}^2 + (kc)^2} . \tag{8.18}$$

In the long-wavelength limit, the relation becomes

$$\omega = \frac{cB}{4\pi ne} k^2 . \tag{8.19}$$

This is the helicon mode of propagation, mentioned earlier in Section 8.1 in connection with the experiment by Bowers et al. (1961).

In the experiment the observed frequency was 32 Hz at B = $10^4$ gauss; with these values Eq. (8.19) yields a half-wavelength of 3.4 mm. This is comparable with the corresponding specimen dimension which is 4 mm. The observed linear dependence of the frequency on the magnetic field is also in agreement with Eq. (8.19).

The left circularly polarized mode may be treated similarly. In place of Eq. (8.14), we now have

$$\left(\frac{kc}{\omega}\right)^2 = 1 - \frac{\omega_{pe}^2}{\omega(\omega + |\Omega_e|)} . \tag{8.20}$$

This relation is depicted by the dashed curve in Fig. 8.5. The cutoff occurs at the frequency

$$\omega_{c\ell} = \frac{1}{2} \left[ (\Omega_e^2 + 4\omega_{pe}^2)^{1/2} - |\Omega_e| \right] . \tag{8.21}$$

No resonance takes place for a left circularly polarized wave in an electron plasma.

As Fig. 8.5 clearly illustrates, a substantial difference exists generally in propagation characteristics between the

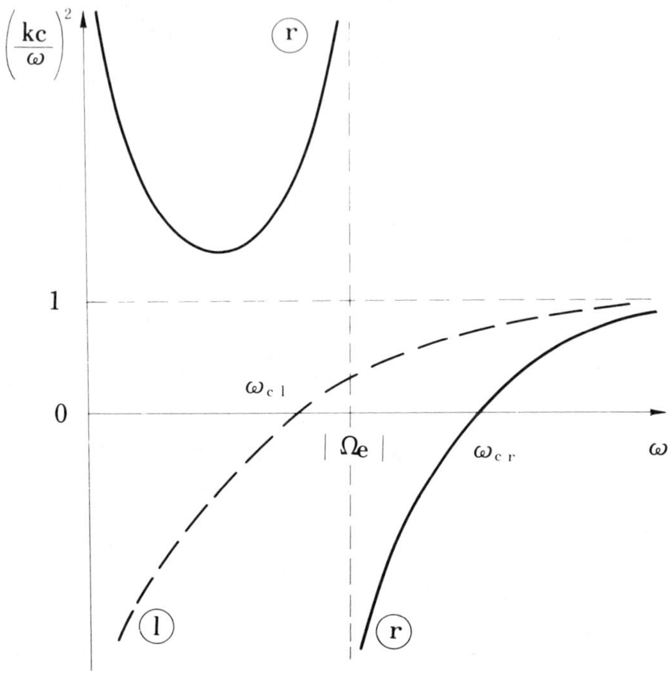


**Figure 8.5.** Right and left circularly polarized waves in an uncompensated electron plasma.

right and left circularly polarized waves in plasmas with a magnetic field. When a linearly polarized wave is injected toward a plasma along the magnetic field, its right and left circularly polarized components propagate differently in the plasma. In those cases where the propagation velocities are different, the plane of linear polarization will rotate after a passage of a finite distance in the plasma, a phenomenon called **Faraday rotation.** In some other cases, only one of the two circularly polarized components may be able to propagate through the plasma; in these circumstance, we have a converter of a linearly polarized wave into a circularly polarized wave.

## 8.5 Compensated Two-Component Plasmas

Let us next consider a two-component plasma consisting of electrons (with subscript e) and holes or positively charged ions (with subscript i); their average charge densities are assumed to be equal.

The dielectric function for the right circularly polarized wave is

$$\varepsilon_r(k, \omega) = 1 - \frac{\omega_{pe}^2}{\omega(\omega - |\Omega_e|)} [1 - W(\frac{\omega - |\Omega_e|}{k(k_B T/m_e)^{1/2}})]$$

$$- \frac{\omega_{pi}^2}{\omega(\omega + \Omega_i)} [1 - W(\frac{\omega + \Omega_i}{k(k_B T/m_i)^{1/2}})] \ . \qquad (8.22)$$

Assuming the conditions for negligible absorption,

$$||\Omega_e| - \omega| \gg k(k_B T/m_e)^{1/2} , \quad \Omega_i + \omega \gg k(k_B T/m_i)^{1/2} , \qquad (8.23)$$

we find the dispersion relation for the right circularly polarized wave

$$(\frac{kc}{\omega})^2 = 1 - \frac{\omega_{pe}^2}{\omega(\omega - |\Omega_e|)} - \frac{\omega_{pi}^2}{\omega(\omega + \Omega_i)} \ . \qquad (8.24)$$

The behavior of this function is illustrated by solid curves in Fig. 8.6.

In the dispersion relation (8.24), the cutoff frequency is located at

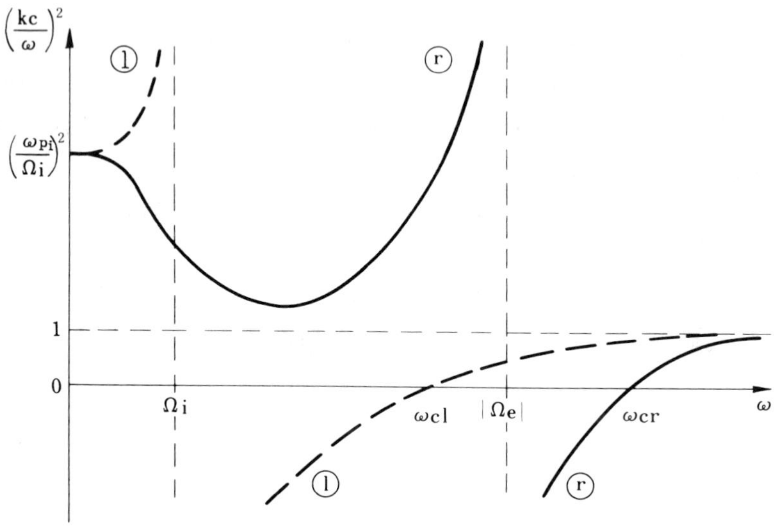


**Figure 8.6.** Right and left circularly polarized waves in a compensated two-component plasma.

$$\omega_{cr} = \frac{1}{2}\,\{[(|\Omega_e| + \Omega_i)^2 + 4(\omega_{pe}{}^2 + \omega_{pi}{}^2)]^{1/2} + |\Omega_e| - \Omega_i\} \tag{8.25}$$

and the resonance frequency is again given by

$$\omega_R = |\Omega_e| \; . \tag{8.26}$$

In the vicinity of this resonance, the first condition of (8.23) becomes violated; strong absorption ensues due to the Dopper-shifted cyclotron resonances of the electrons. The second condition of (8.23), on the other hand, is trivially satisfied in ordinary circumstances.

The feature to be remarked here is again the appearance of a low-frequency propagation mode. Neglecting the first term on the right-hand side of Eq. (8.24), we find in the low-frequency domain

$$\omega = \frac{(|\Omega_e| - \Omega_i)(kc)^2}{\omega_{pe}^2 + \omega_{pi}^2 + (kc)^2} \left\{\frac{1}{2} + \left(\frac{|\Omega_e|\Omega_i[\omega_{pe}^2 + \omega_{pi}^2 + (kc)^2]}{(|\Omega_e| - \Omega_i)^2(kc)^2} + \frac{1}{4}\right)^{1/2}\right\} . \tag{8.27}$$

In this derivation we use the condition for carrier-charge compensation

$$|\Omega_e|\omega_{pi}^2 = \Omega_i\omega_{pe}^2 . \tag{8.28}$$

In the long-wavelength limit such that

$$(kc)^2 << 4\omega_{pi}^2 << \omega_{pe}^2 , \tag{8.29}$$

Eq. (8.27) approaches

$$\omega = (\Omega_i/\omega_{pi})kc . \tag{8.30}$$

The mode described by Eq. (8.30) is the Alfvén wave; its propagation velocity

$$c_A = \frac{B}{(4\pi n_i m_i)^{1/2}} = \frac{B}{(4\pi\rho_m)^{1/2}} \tag{8.31}$$

is called the Alfvén velocity; in Eq. (8.31), $\rho_m = n_i m_i$ refers to the mass density of the plasma.

We now return to the interference experiment by Williams and Smith (1964), described earlier in Section 8.1. Since the refractive index is given by $\eta = c/c_A$, the fringe index plotted in Fig. 8.4 is expressed as

$$N = (A/B)\rho_m^{1/2} \tag{8.32}$$

where A contains all of the constants. Therefore one expects that the fringe index for the Alfvén wave should be a straight line plotted as a function of 1/B, a feature clearly observed in Fig. 8.4. It was also found in the experiment that effective refractive index had no frequency dependence, and thus true Alfvén wave behavior was demonstrated.

Propagation of the left circularly polarized wave may be analyzed analogously. Assuming

$$|\Omega_i - \omega| >> k(k_B T/m_i)^{1/2}, \qquad |\Omega_e| + \omega >> k(k_B T/m_e)^{1/2} \tag{8.33}$$

we obtain the dispersion relation

$$\left(\frac{kc}{\omega}\right)^2 = 1 - \frac{\omega_{pe}^2}{\omega(\omega + |\Omega_e|)} - \frac{\omega_{pi}^2}{\omega(\omega - \Omega_1)} \; . \tag{8.34}$$

This relation is illustrated by the dashed curves in Fig. 8.6. The cutoff frequency is calculated as

$$\omega_{c\ell} = \frac{1}{2} \{[(|\Omega_e| + \Omega_i)^2 + 4(\omega_{pe}^2 + \omega_{pi}^2)]^{1/2} - |\Omega_e| + \Omega_i\} \tag{8.35}$$

and the resonance frequency is

$$\omega_L = \Omega_i \; . \tag{8.36}$$

In the vicinity of this resonance, the first condition of (8.33) becomes violated; the Doppler-shifted cyclotron resonances of the ions now cause strong attenuation to the wave.

In the low-frequency domain, Eq. (8.34) becomes

$$\omega = \frac{(|\Omega_e| - \Omega_i)(kc)^2}{\omega_{pe}^2 + \omega_{pi}^2 + (kc)^2} \left\{ -\frac{1}{2} + \left( \frac{|\Omega_e|\Omega_i[\omega_{pe}^2 + \omega_{pi}^2 + (kc)^2]}{(|\Omega_e| - \Omega_i)^2 (kc)^2} + \frac{1}{4} \right)^{1/2} \right\} . \tag{8.37}$$

This dispersion relation reduces to that of the Alfvén wave, Eq. (8.30), in the long-wavelength limit $(kc)^2 \ll 4\omega_{pi}^2$. The Alfvén wave can, therefore, propagate as either a right or a left circularly polarized wave, and hence also as a linearly polarized wave, in a compensated magnetized plasma.

# Chapter 9

# Nonlinear Effects

When an external disturbance is applied to a plasma, physical quantities such as the electric charge distribution exhibit transient variations and deviations away from their equilibrium values. When the magnitude of such a deviation is reasonably small one can analyze and understand the transient behavior through a linearized approximation. When the amplitude of variation is large, however, nonlinear effects become prominent and give rise to new features in the transient variation. In this chapter we take up some of those nonlinear effects and consider relevant experimental indications.

## 9.1 Trapped Particles

We consider a variation of the electrostatic potential

$$\phi(x, t) = \phi_0 \cos(kx - \omega t) \tag{9.1}$$

propagating in the x direction with wave number k and frequency $\omega$ in a plasma. Transforming from the laboratory system (x, t) to the coordinate system (X, t) comoving with the wave, one has a relation

$$X = x - (\omega/k)t \ , \tag{9.2}$$

so that Eq. (9.1) becomes

$$\phi(X, t) = \phi_0 \cos(kX) \ . \tag{9.3}$$

The x component of the particle velocity v in the plasma likewise transforms into

$$V = v - (\omega/k) \tag{9.4}$$

in the new coordinates; the total energy of the particle (mass M, charge Ze) is given by

$$W = \frac{1}{2} MV^2 + Ze \ \phi_0 \cos(kX) \ . \tag{9.5}$$

As Fig. 9.1 illustrates, the motion of a plasma particle follows a trajectory W = constant in the phase space (X, V). A particle with $W < 0$ is trapped in a trough of potential and performs a periodic motion along a closed contour in the phase space. Such a particle is called a **trapped particle.**

Analogously to the case of the motion of a pendulum, those particles trapped near the bottom of the potential trough follow an equation of motion of a harmonic-oscillator type. The electric field associated with Eq. (9.3) is given by $\phi_0 k \sin(kX)$, so that the equation of motion for $|kX| \ll 1$ is

$$M\ddot{X} = Ze \ \phi_0 k^2 X \ .$$

Hence the frequency of the periodic motion is given by

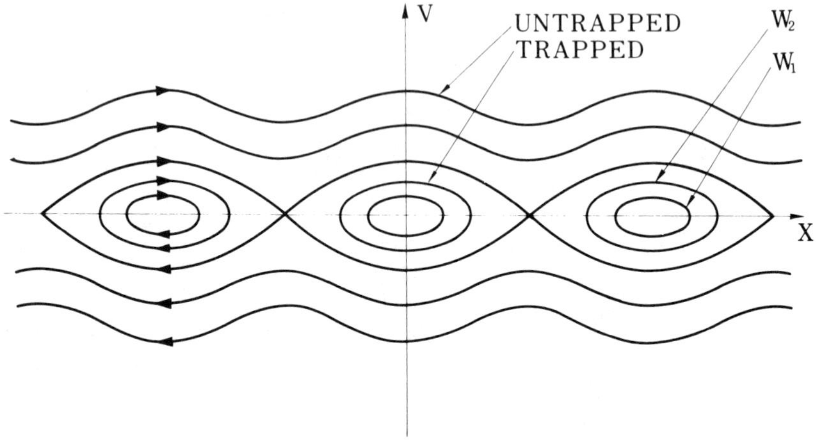


**Figure 9.1.** Trajectory of trapped and untrapped particles in the phase space. Phase evolutions of the trapped particles $W_1$ and $W_2$ are slightly different.

$$\omega_B = |Ze\ \phi_0 k^2/M|^{1/2} \tag{9.6}$$

which we shall call the **bounce frequency.**

Such existence of the trapped particles and the associated periodic motion are consequences of the wave-particle interaction. In Section 2.7 we treated certain aspects of the wave-particle interaction and thereby derived the expression for the Landau damping as given by Eq. (2.41). In the derivation we considered linear effects only and did not take into account those effects arising from the periodic motion of the trapped particles. In order for the Landau damping to become a major effect in the wave-particle interaction, the wave amplitude must be sufficiently small so that the condition

$$\omega_B \ll |\gamma_k| \tag{9.7}$$

is satisfied. The condition (9.7) implies that the wave amplitude decays before a trapped particle completes one cycle of its periodic motion; the effect of the trapped oscillation would not appear.

## 9.2 The Amplitude Oscillation

Let us then consider the kind of behavior expected when the wave amplitude is sufficiently large, so that

$$\omega_B \gg |\gamma_k| \tag{9.8}$$

is satisfied. We consider this problem along with Fig. 9.2, which is analogous to Fig. 2.7. For the one-dimensional velocity distribution at t = 0 as depicted by a solid curve, the gradient at $v = \omega/k$ is negative and thus the wave starts to decay in accordance with Eq. (2.41). After a half period ($t = \pi/\omega_B$) of the trapped oscillation, those particles trapped in the shaded area of Fig. 9.2 assume a distribution shown by the dashed curve, which is nearly an inversion of the solid curve with respect to $v = \omega/k$. Since some of the trapped particles do not completely obey the harmonic equation of motion, the inverted distribution contains a certain degree of

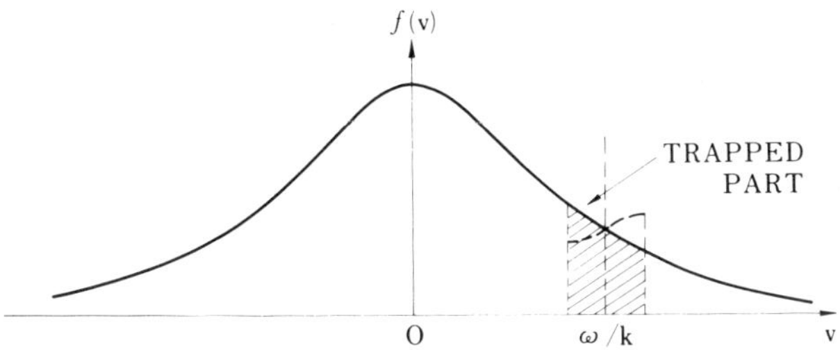


**Figure 9.2.** Velocity distribution of the particles and the trapped part.

"smoothing." We, nonetheless, expect a positive value for $\partial f_1(v)/\partial v|_{v=\omega/k}$ at $t = \pi/\omega_B$; wave amplitude would start to grow according to Eq. (2.41). At $t = 2\pi/\omega_B$, another half period later, we expect that most of the trapped particles would return to the state at $t = 0$ in the phase space; $\partial f_1(v)/\partial v|_{v=\omega/k}$ would again assume a negative value and the wave would tend to decay.

Those considerations indicate that the rate of Landau damping Eq. (2.41) may be subjected to a temporal modulation, as illustrated in Fig. 9.3, owing to the oscillatory behavior of the trapped particles in the phase space. Since $\gamma_k$ represents the decay rate of the wave amplitude, one finds an equation

$$\frac{d|\phi_0|^2}{dt} = -2\gamma_k(t)|\phi_0|^2 ,$$

whose solution yields

$$|\phi_0(t)|^2 = |\phi_0(0)|^2 \exp[-2 \int_0^t dt'\gamma_k(t')] . \tag{9.9}$$

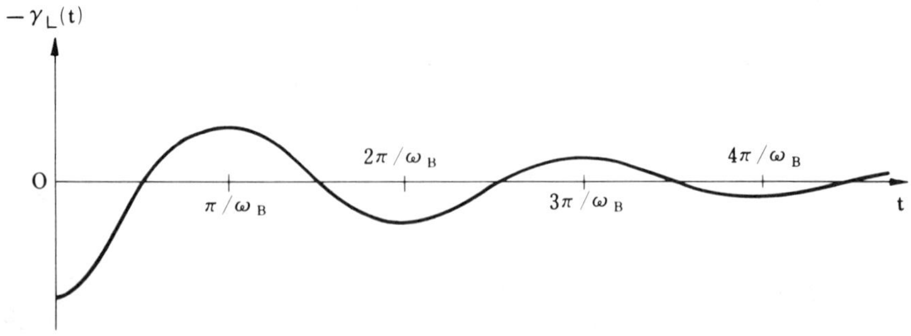


**Figure 9.3.** Schematic behavior of the Landau decay rate $\gamma_L$.

In light of Fig. 9.3, we find that the amplitude itself would oscillate at a frequency in the vicinity of $\omega_B$.

Such a nonlinear effect expected for a large amplitude plasma wave is called the **amplitude oscillation.** The amplitude oscillation of the plasma wave was discovered experimentally first by Malmberg and Wharton (1967). The experiment was performed by using an interferometric technique as illustrated in Fig. 3.1. The electron density of the plasma was approximately $5 \times 10^8$ cm$^{-3}$, the electrons assumed a Maxwellian distribution with temperature of approximately 7 eV, and the associated Debye length was approximately 0.1 cm. The mean free path of an electron was said to be extremely large; for electron-ion collisions it was estimated to be approximately $10^5$ cm and for electron-neutral collisions, $4 \times 10^3$ cm.

In Fig. 3.1 one applies an oscillatory potential to the transmitter grid (A) and excites the plasma wave. The wavelength of the plasma wave so excited was measured to be 2.3 cm. The amplitude detected at the receiver grid (B) is observed as a function of distance between grids (A) and (B); the results are shown in Fig. 9.4. The intensity of the receiver signal increases as we move from the data A, B, to C.

As may be clear from the foregoing arguments, the appearance of amplitude oscillation in the plasma wave depends on comparison between the magnitudes of $\omega_B$ and $\gamma_k(0)$. In the experiment pertaining to Fig. 9.4, however, the spatial propagation, rather than the temporal propagation, of the large-amplitude wave has been measured. As remarked in Section 3.1, it is more relevant to work with a complex wave number solution, $k_r(\omega) + ik_i(\omega)$, than a complex frequency solution, $\omega_k - i\gamma_k$, for the propagation of a small-amplitude plasma wave. The condition for the amplitude oscillations (Malmberg and Wharton 1967) is then given by

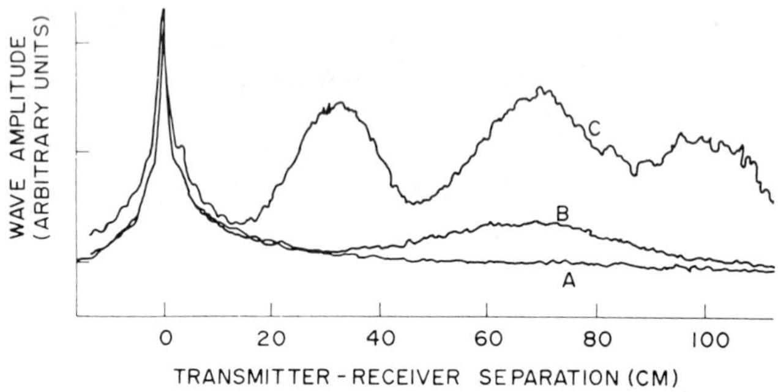


**Figure 9.4.** Wave amplitude vs. position. Transmitter voltage 0.9, 2.85, and 9 V, for curves A, B, and C, respectively. After Malmberg and Wharton (1967).

$$k_{osc}/k_i(\omega) \gtrsim (2\pi)^{1/2}$$

where $k_{osc} = \omega_B k_r(\omega)/\omega$ is the wave number for the spatial variation of the amplitude oscillation. For case A in Fig. 9.4, it has been estimated that $k_{osc}/k_i(\omega) = 2.3$; an amplitude oscillation is not expected. For case B, $k_{osc}/k_i(\omega) = 4.1$; the amplitude oscillation is marginally expected. Finally for case C, $k_{osc}/k_i(\omega) = 7.4$; the observed amplitude oscillation is well expected theoretically. We may regard those as an experimental demonstration of the amplitude oscillation effect brought about by the trapped particles.

The wave number $k_{osc}$ of the spatial variation of the amplitude oscillation is expected to be proportional to the square root of the amplitude $\phi_0$ of the plasma wave. This feature can be examined experimentally through a comparison between the signal strength of the transmitter and the wavelength of the spatial oscillation; the behavior in accordance with Eq. (9.6) has been confirmed.

## 9.3 Ion-Acoustic Wave Solitons

We next consider a nonlinear effect associated with a flow of the plasma.

In Fig. 9.5 we show an experimental result on the ion-acoustic wave soliton performed by Ikezi (1973). The plasma was produced in Ar gas at the pressure (1 - 3) $\times 10^{-4}$ Torr, electron density $n \simeq 10^9 - 10^{10}$ cm$^{-3}$, electron temperature $T_e \simeq$ 3 eV, and ion temperature $T_i < 0.2$ eV. The actual experimental setup was substantially different from that schematically shown in Fig. 3.1; the general idea of the experiment can still be looked upon as a measurement of space-time propagation of a density wave. The potential at the grid (A) in Fig. 3.1 is suddenly changed (the top trace of Fig. 9.5), and a local variation of plasma density is produced. The way in which such a variation propagates in the plasma is measured by a probe situated at a distance x away. The successive traces in Fig. 9.5 are those signals representing the temporal variations of electron density measured at various values of x.

A number of salient features are noteworthy in those signals. First the front edge of the variation steepens as the

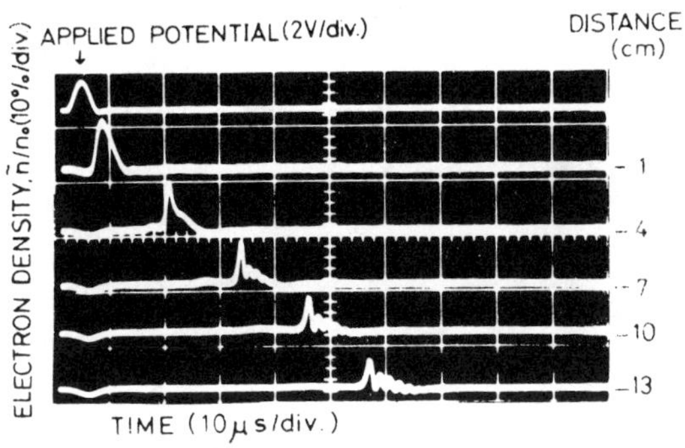


**Figure 9.5.** Ion-acoustic wave solitons. After Ikezi (1973).

wave propagates. As the wave steepens, several peaks are then produced out of the original wave; those peaks are arranged successively according to the peak amplitude. These are the **solitons** associated with the ion-acoustic wave.

As we mentioned in Section 4.8, the ion-acoustic wave is the collective motion of the ions; many ions participate cooperatively in its excitation. For the analysis of the nonlinear effect relevant to this experiment, we may thus regard the ion component in the plasma as a kind of fluid. Returning to the notation of Section 4.4, we distinguish between the quantities associated with electron and ion by the suffixes e and i. Denoting by $u_i$ the flow velocity of the ion fluid, we choose the x axis in this direction and assume that various physical quantities are functions of only x and time t.

For the surfaces 1 and 2 at x and x + dx in the plasma, as shown in Fig. 9.6, we calculate the rate of ion-density vari-

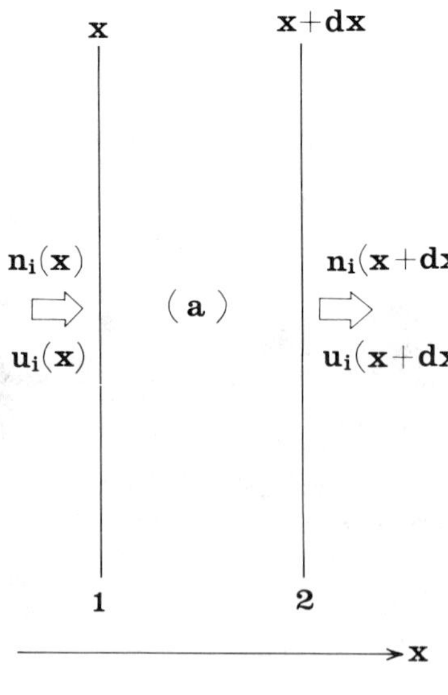


**Figure 9.6.** Variation of ionic fluid.

ation in the space (a) between them. The ions moving into the space through surface 1 per unit time, per unit area is $n_i(x)u_i(x)$; similarly the ions moving out through surface 2 is $n_i(x + dx)u_i(x + dx)$. The difference between the two contributes to the variation of the ion density in the space (a)

$$\frac{\partial n_i}{\partial t} dx = n_i(x)u_i(x) - n_i(x + dx)u_i(x + dx) \ .$$

Expanding the right-hand side with respect to dx and retaining the first-order terms only, we obtain

$$\frac{\partial n_i}{\partial t} + \frac{\partial}{\partial x}(n_i u_i) = 0 \ . \tag{9.10}$$

This is the **continuity equation** for the ion fluid.

We next calculate the rate $\partial(m_i n_i u_i)/\partial t$ of the change in the momentum density associated with the ions in the space (a). The amount of momenta carried by those ions flowing through the surface 1 per unit time, per unit area is $m_i n_i(x)u_i(x)^2$; similarly, the amount moving out through the surface 2 is $m_i n_i(x + dx)u_i(x + dx)^2$. In addition, the amount of momentum changes in accordance with Newton's law when a force is exerted on the ions in (a). In the presence of an electrostatic potential $\phi$ in the plasma, the force per ion ($Z = 1$ is assumed) is $-e\partial\phi/\partial x$. Hence

$$\frac{\partial(m_i n_i u_i)}{\partial t} dx = m_i n_i(x)u_i(x)^2 - m_i n_i(x + dx)u_i(x + dx)^2 - e n_i \frac{\partial\phi}{\partial x} dx \ .$$

Again from the first-order terms with respect to dx we find

$$\frac{\partial(n_i u_i)}{\partial t} + \frac{\partial}{\partial x}(n_i u_i^2) = -\frac{e}{m_i} n_i \frac{\partial \phi}{\partial x} . \tag{9.11}$$

With the aid of Eq. (9.10), we thus obtain

$$\frac{\partial u_i}{\partial t} + u_i \frac{\partial u_i}{\partial x} = -\frac{e}{m_i} \frac{\partial \phi}{\partial x} . \tag{9.12}$$

In the derivation of Eq. (9.11) the contributions of viscosity and the partial pressure of ions are neglected. The partial pressure of the ions is proportional to the ion temperature $T_i$; as we examined in Section 4.6, $T_e \gg T_i$ must be satisfied for the existence of the ion-acoustic wave. Under these circumstances the effect of the partial pressure of ions can be ignored. Since we are dealing with a weakly coupled plasma, the viscosity term arising from interparticle collisions can likewise be ignored in light of the discussion in Section 2.8.

The ion-acoustic wave is a low-frequency oscillation from the viewpoint of the electrons; one can express the electron density $n_e$ in the quasi-static approximation

$$n_e = n \exp\left(\frac{e\phi}{k_B T}\right) \tag{9.13}$$

as explained in Section 2.2. Here n refers to the average number density of electrons or ions. One can thus close the set of equations by Eqs. (9.10), (9.12), (9.13) and the Poisson equation

$$\frac{\partial^2 \phi}{\partial x^2} = 4\pi e(n_e - n_i) , \tag{9.14}$$

which determines the electrostatic potential.

That the collective motion described by this set of equations corresponds to the ion-acoustic wave discussed in Section 4.6 can be confirmed in the following way: We write the electron and ion densities as sums of their average value n and deviations thereof, $\delta n_e$ and $\delta n_i$. We then linearize the set of equations (9.10), (9.12), (9.13) and (9.14) with respect to variations $\delta n_e$, $\delta n_i$, $\phi$, and $u_i$. Assuming that those variations have a sinusoidal space-time variation $\cos(kx - \omega t)$, we arrive at the relation (4.21) between k and $\omega$.

In the confines of the linear analysis mentioned above, the second term on the left-hand side of Eq. (9.12) plays no parts. This term, being a second-order nonlinear term with respect to the variation $u_i$, becomes significant at a large wave amplitude. The steepening of the wave observed in Fig. 9.5 results from this nonlinear effect brought about by the plasma flow.

To extract the relevant nonlinear effect, we consider an equation

$$\frac{\partial u_i}{\partial t} + u_i \frac{\partial u_i}{\partial x} = 0 \tag{9.15}$$

where the Coulomb-force term on the right-hand side of Eq. (9.12) is neglected. Let us note that the "coefficient" $u_i$ in the second term on the left-hand side plays the part of "propagation velocity" for the variation. Assume at time $t_1$ that $u_i$ and $n_i$ have distributions as shown by the dashed curves in Fig. 9.7. The portion where $u_i$ is large moves fast in accordance with Eq. (9.15); at $t_2(> t_1)$ the distributions will become as depicted by solid curves in Fig. 9.7. As a consequence the wave front of $u_i$ steepens sharply; a spatial variation of density is induced according to the continuity equation (9.10).

As the wave front of the ion-acoustic wave steepens and the

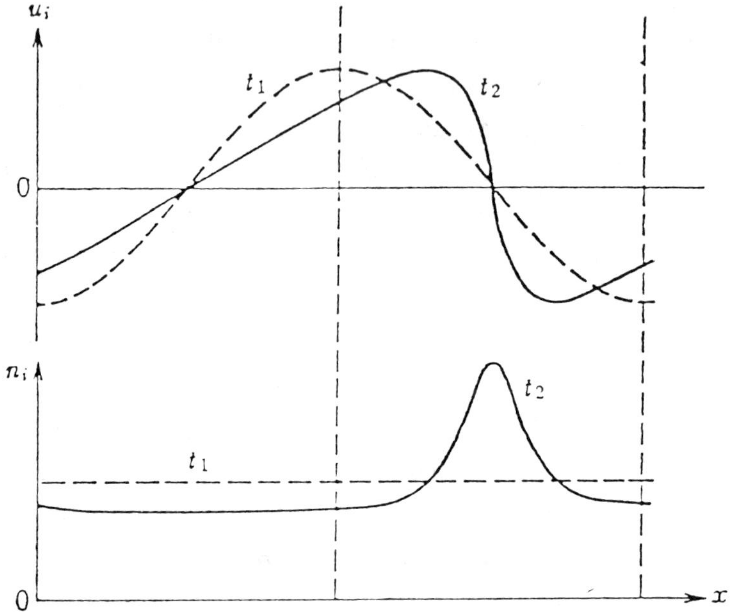


**Figure 9.7.** Steepening of wave front due to the nonlinear effect of plasma flow.

variations of density and flow velocity become abrupt spatially, a dispersion effect of the wave begins to take effect. In the long-wavelength domain, the ion-acoustic wave obeys the dispersion relation (4.22) of the acoustic type; no change of the wave form will result along the propagation of the wave. When the steepening of the wave front takes place owing to the nonlinear effect, the short-wavelength components of the variation increase. Those obey the propagation characteristics in accord with Eq. (4.21); the phase and group velocities are now functions of the wavelength. A change of the wave form results as a consequence of this dispersion effect. The solitons observed in Fig. 9.5 are said to be generated through balance between the steepening effect of nonlinearity and the dispersion effect in the short-wavelength domain.

One can in fact derive*

$$\frac{1}{s}\frac{\partial\Phi}{\partial t} + \Phi\frac{\partial\Phi}{\partial\xi} + \frac{1}{2k_{De}^2}\frac{\partial^3\Phi}{\partial\xi^3} = 0 \tag{9.16}$$

approximately starting with Eqs. (9.10), (9.12), (9.13) and (9.14) (Washimi and Taniuti 1966). Here

$$\Phi = \frac{e\phi}{k_B T} \tag{9.17}$$

$$\xi = x - st \ . \tag{9.18}$$

The second term on the left-hand side of Eq. (9.16) corresponds to the second term on the left-hand side of Eq. (9.12), the nonlinear effect of the plasma flow. The third term on the left-hand side of Eq. (9.16) represents the dispersion effect, derived from the second term on the right-hand side of the expansion

$$\omega_k = sk\left(1 - \frac{k^2}{2k_{De}^2} + \ldots\right) \tag{9.19}$$

---

*A simple way of deriving Eq. (9.16) is the following. In addition to Eq. (9.17), we introduce dimensionless variables $n'$ and $u'$ through $n_i = n(1 + n')$ and $u_i = su'$. We then carry out coordinate transformation via Eq. (9.18), and expand Eqs. (9.10), (9.12), (9.13) and (9.14) up to the quadratic terms with respect to the perturbations. The equations are combined and simplified because we may take $\Phi = n' = u'$ in the linear approximation; the result then is Eq. (9.16).

of Eq. (4.21) with respect to $k^2$. The principal term sk of Eq. (9.19) is taken care of by the transformation Eq. (9.18).

Equation (9.16) takes the form of the **Korteweg-de Vries equation** derived in 1895 on the shallow-water wave; it has a special solution representing a soliton

$$\Phi = \Phi_0 \mathrm{sech}^2\left(\frac{x - ut}{\Delta}\right) \tag{9.20}$$

$$u = s\left(1 + \frac{\Phi_0}{3}\right) \tag{9.21}$$

$$\Delta = \frac{1}{k_{De}} \sqrt{\frac{6}{\Phi_0}} \ . \tag{9.22}$$

According to this solution the soliton exists only in the positive direction of the potential $\phi = (k_B T/e)\Phi$; the propagation velocity u increases from s and the width $\Delta$ decreases as the amplitude $\phi_0$ increases. These features are also exhibited in the experimental result of Fig. 9.5.

## 9.4 Ponderomotive Force

Presence of a longitudinal wave or a transverse wave in a plasma implies an energy density proportional to the square of the wave amplitude. When this energy density varies spatially, a force proportional to its gradient is exerted onto the plasma. Generally such a force is referred to as a **ponderomotive force**; it is a nonlinear effect essential in the description of large amplitude waves in a plasma.

To understand existence of the ponderomotive force, we consider an example illustrated in Fig. 9.8. An electric field $\vec{E}$ in the x direction is applied to a dielectric medium, which contains n "molecules" per unit volume; the field strength is assumed to vary slowly (relative to the molecular size) in the x direction. If a dipole moment $\vec{d} = e\Delta\vec{r}$ is induced in a mole-

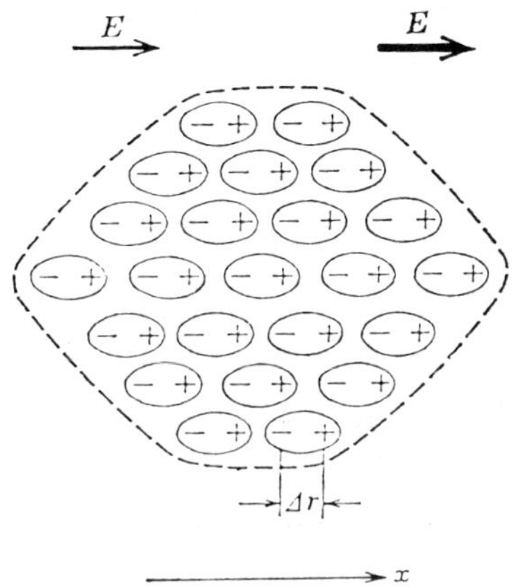


**Figure 9.8.** Ponderomotive force acting on a dielectric medium.

cule under the action of the electric field, then the force on the molecule is

$$\vec{f} = e[\vec{E}(\vec{r} + \Delta\vec{r}) - \vec{E}(\vec{r})] = (\vec{d} \cdot \frac{\partial}{\partial \vec{r}})\vec{E} \quad . \tag{9.23}$$

The force is thus given by a derivative of the vector field $\vec{E}$ in the $\vec{d}$ direction.

Denoting the molecular polarizability by $\mu$, one has $\vec{d} = \mu\vec{E}$, so that Eq. (9.23) becomes

$$\vec{f} = \mu(\vec{E} \cdot \frac{\partial}{\partial \vec{r}})\vec{E} \quad . \tag{9.24}$$

For the configuration of Fig. 9.8, this force is exerted in the x direction with a magnitude expressed as

$$f_x = \frac{\mu}{2} \frac{\partial}{\partial x} (E^2) \quad . \tag{9.25}$$

The force acting on the medium per unit volume is then calculated from Eq. (9.25) as

$$F_x = nf_x = \frac{n\mu}{2}\frac{\partial}{\partial x}(E^2) \quad .$$

Since $n\mu = (\varepsilon - 1)/4\pi$ where $\varepsilon$ is the dielectric constant of the medium, one obtains

$$F_x = \frac{\varepsilon - 1}{8\pi}\frac{\partial}{\partial x}(E^2) \quad . \tag{9.26}$$

It is important here to note that the direction of this force depends on the sign of $\varepsilon - 1$.

We have thus calculated the force acting on a dielectric medium when a static electric field varies slowly in space. A force analogous to it arises in the presence of a high-frequency electric field associated with, for example, the plasma oscillation. Let us assume an oscillatory electric field (longitudinal wave) in a plasma with frequency $\omega$ and wave number k, whose amplitude varies slowly in the x direction. The field component in the x direction is expressed as

$$E(x, t) = E_L(x)\cos(kx - \omega t + \phi) \tag{9.27}$$

where we assume that $E_L(x)$ remains almost constant over a distance of the wavelength $2\pi/k$.

In the case of the high-frequency electric field (9.27), we replace the dielectric constant $\varepsilon$ in Eq. (9.26) with the dielectric function $\varepsilon(\vec{k}, \omega)$, and $E^2$ with an average of the square of $E(x, t)$. This average is carried out over the high-frequency component, so that

$$\langle E(x, t)^2 \rangle = E_L{}^2(x)/2 \quad . \tag{9.28}$$

Hence, the slow spatial variation contained in $E_L(x)$ remains after the average. For a high-frequency oscillation, the expansion of Eq. (A5.20) is applicable. Retaining up to the terms proportional to $\omega^{-2}$, we find an expression for the ponderomotive force of plasma oscillation acting on the plasma as

$$F_x = - \frac{\omega_p^2}{16\pi\omega^2} \frac{\partial}{\partial x} E_L^2(x) \quad . \tag{9.29}$$

According to Eq. (9.29), the ponderomotive force acts from the region of large amplitude plasma wave to the region of small amplitude plasma wave; the plasma is thus pushed away from the domain of the large amplitude wave. As the plasma density decreases, the local frequency of the plasma oscillation also decreases. The plasma oscillation is consequently confined in a domain where plasma density is "cavitated," as Fig. 9.9 illustrates. **Self-trapping** of a large amplitude plasma wave thus takes place owing to the action of the ponderomotive force and subsequent localization of the plasma wave (Zakharov 1972;

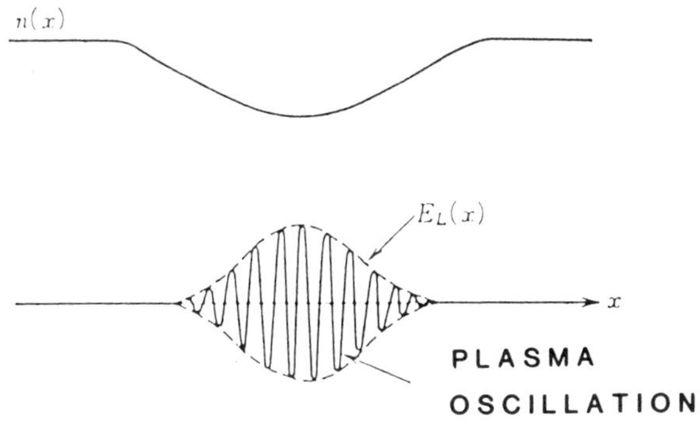


**Figure 9.9.** Self-trapping of large-amplitude plasma oscillations and cavitation of the plasma.

Kim, Stenzel, and Wong 1974, Wong and Cheung 1984).

A ponderomotive force operates as a nonlinear effect also when a large amplitude electromagnetic wave (transverse wave) propagates in a plasma. As Eq. (9.24) indicates, the ponderomotive force is mainly operative in the direction of the oscillating electric field; for the transverse wave it is perpendicular to the propagation. An electromagnetic field propagating in the x direction with the electric field

$$E_y(\vec{r}, t) = E_T(\vec{r})\cos(kx - \omega t + \phi) \tag{9.30}$$

directed in the y direction, the ponderomotive force in the y direction is given by

$$F_y = \frac{\varepsilon_T(\omega) - 1}{16\pi} \frac{\partial}{\partial y} (E_T{}^2) \quad . \tag{9.31}$$

Here $\varepsilon_T(\omega)$ is the dielectric constant of the plasma against the electromagnetic wave as considered in Sections 4.1 and 8.3.

Since the directional dependence in the case of a transverse wave is different from that in the case of the longitudinal wave, a nonlinear effect associated with the ponderomotive force appears in the form of **self-focusing.**

Consider a (cylindrical) beam of electromagnetic wave injected in a plasma (see Fig. 9.10). When $\varepsilon_T(\omega) - 1 > 0$, the plasma is attracted to the domain of strong beam in accordance with Eq. (9.31). As the plasma density increases, $\varepsilon_T(\omega) - 1$ further increases [see e.g., Eq. (8.30)]. With an increase of $\varepsilon_T(\omega)$ the phase velocity of the wave in this domain is reduced in view of Eq. (4.9); the wave front of the electromagnetic wave converges toward this region. As the electromagnetic wave converges, $E_T{}^2$ increases and plasma is further attracted.

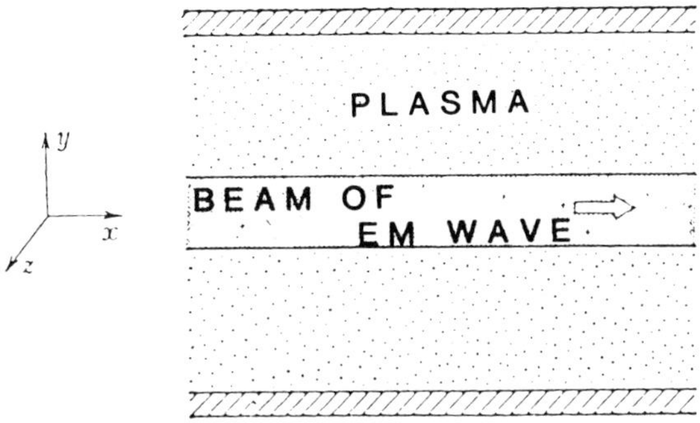


**Figure 9.10.** Injection of a beam of electromagnetic wave into a plasma.

The self-focusing proceeds through these processes. In these circumstances, plasma is attracted in the domain of the strong beam as Fig. 9.11 illustrates.

When $\varepsilon_T(\omega) - 1 < 0$, a self-focusing of the beam also takes place. In this case plasma is rejected from the domain of the strong beam in accordance with Eq. (9.31). As the plasma density decreases $|\varepsilon_T(\omega) - 1|$ also decreases but $\varepsilon_T(\omega)$ itself increases [see e.g., Eq. (4.10)]. Consequently a convergence of the wave front takes place following a process analogous to that mentioned above; an increase of $E_T{}^2$ acts to decrease the plasma density locally and thus a self-focusing proceeds. In these circumstances plasma is rejected from the domain of the strong beam as Fig. 9.11 illustrates.

The ponderomotive force effects and particularly the self-trapping processes as described in Fig. 9.9 play important parts in the formulation of a strong-turbulence theory for plasma waves. For example, instability of self-focused light beams has been investigated by Zakharov (1967): Solutions describing stationary one-dimensional propogation of a coupled

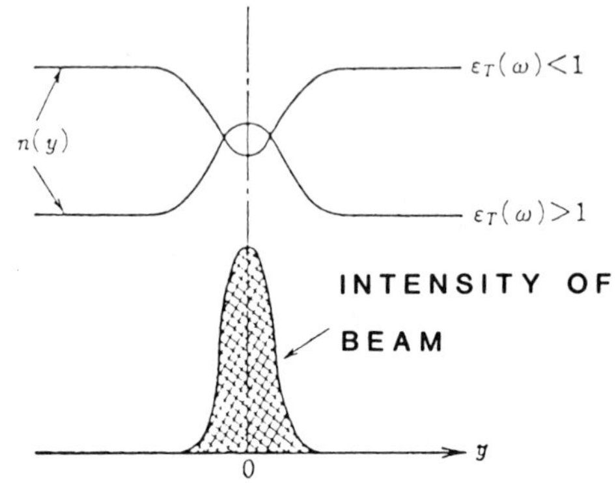


**Figure 9.11.** Self-focusing of a beam of electromagnetic wave injected into a plasma.

nonlinear electron-plasma wave, and a nonlinear ion-acoustic wave have been obtained by Nishikawa, Hojo, Mima, and Ikeji (1974). The reader is referred to a review on recent developments in those fields given by Goldman (1984).

## Chapter 10

# High-Density Plasma in Stellar Interior

Typical examples of a strongly coupled classical plasma may be seen in the system of ions inside a highly evolved star such as a white dwarf, and in the outer crust of a neutron star. Owing to the strong Coulomb coupling, the interionic correlations exhibit features markedly different from those observed in a weakly coupled plasma. Those features significantly affect the basic properties of the system, such as the equation of state, the rate of elementary processes, and the transport coefficients. In this chapter we investigate some of the problems associated with strong correlations in dense plasmas.

## 10.1 Strongly Coupled Classical Plasmas

Consider stellar matter with mass density $\rho_m$ at temperature T, consisting of a nuclei with mass number A and the charge number Z. We are here interested in the parameter domain: $10^4 \leq \rho_m \leq 10^9$ [g/cm$^3$] and $10^6 \leq T \leq 10^8$ [K]. For a carbon plasma, a typical constituent of a white dwarf, A = 12 and Z = 6.

Assuming that each nucleus contributes Z conduction electrons, we compute the number density of the electrons as

$$n_e = 6.0 \times 10^{29} \left(\frac{Z}{A}\right)\left(\frac{\rho_m}{10^6 \text{ g/cm}^3}\right) \quad [\text{cm}^{-3}] \ . \tag{10.1}$$

The $r_s$ parameter of Eq. (1.7) is then given by

$$r_s = 1.4 \times 10^{-2} \left[\left(\frac{Z}{A}\right)\left(\frac{\rho_m}{10^6 \text{ g/cm}}\right)\right]^{-1/3} \ . \tag{10.2}$$

With the relativistic correction, the Fermi energy of the electrons is

$$E_F = mc^2\{[1 + 1.96 \times 10^{-4} \ r_s^{-2}]^{1/2} - 1\} \tag{10.3}$$

where m is the rest mass of an electron.

The condition that the atomic nuclei are all stripped of their orbital electrons may be derived from the requirement that the Fermi energy be greater than the binding energy of an orbital electron, $E_F > 13.6 \ Z^2$ [eV], that is

$$\rho_m > 0.38 \ A \ Z^2 \quad [\text{g/cm}^3] \ . \tag{10.4}$$

When this condition is not satisfied, the atomic nuclei may retain some of the orbital electrons.

Equation (10.2) implies that $r_s < 0.1$ in the parameter domain of the present interest. The electron system thus constitutes a weakly coupled ($r_s \ll 1$) degenerate plasma with an immensely large Fermi energy ($E_F \approx mc^2$). Since $r_s \ll 1$, the polarizability of the electron system may be negligible; for, in Eq. (6.16), $\varepsilon(\vec{k}, \omega) - 1$ is proportional to $r_s$. The electron system makes an ideal neutralizing background of negative charges to the ion system, which constitutes a strongly coupled classical plasma.

To see this, we compare between the thermal de Broglie wavelength $\hbar(Mk_BT)^{-1/2}$ of the ions and the ion-sphere radius, Eq. (1.5), where M is the ionic mass. Numerically we find

$$\frac{a}{\hbar(Mk_BT)^{-1/2}} \simeq 12\left(\frac{A}{12}\right)^{5/6}\left(\frac{\rho_m}{10^6\ \mathrm{g/cm^3}}\right)^{-1/3}\left(\frac{T}{10^7\ \mathrm{K}}\right)^{1/2} . \quad (10.5)$$

When this ratio is much greater than unity we may ignore the quantum-mechanical interference effects between ions.

The Coulomb coupling constant, Eq. (1.10), for the classical plasma is then calculated as

$$\Gamma = 35.9\ \left(\frac{Z}{6}\right)^2\left(\frac{A}{12}\right)^{-1/3}\left(\frac{\rho_m}{10^6\ \mathrm{g/cm^3}}\right)^{1/3}\left(\frac{T}{10^7\ \mathrm{K}}\right)^{-1} . \quad (10.6)$$

In the interior of a highly evolved star, this coupling constant is usually greater than unity; such a dense stellar matter thus makes a good example of a strongly coupled classical plasma.

## 10.2 Monte Carlo Simulation Results

In their pioneering work, Brush, Sahlin, and Teller (1966) performed Monte Carlo simulation for the classical one-component plasma* (OCP) based on the algorithm of Metropolis et al. (1953). Several years later, Hansen (1973) carried out similar numerical experiments with improved accuracy. Recently, Slattery, Doolen, and DeWitt (1980, 1982) obtained the Monte Carlo simulation data with further improvement of the accuracy.

Figure 10.1 shows the radial distribution function g(r) com-

---

* See Section 1.2 for the definition.

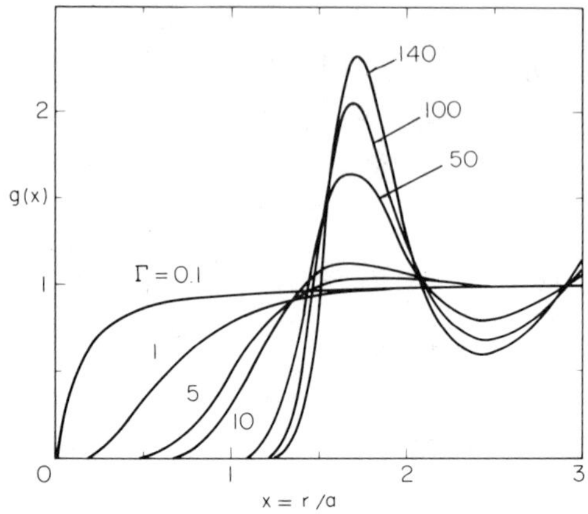


**Figure 10.1.** The radial distribution function for OCP at various values of Γ, where a refers to the ion-sphere radius Eq. (1.5).

puted through these experiments for various values of the OCP coupling constant, Eq. (1.10). The radial distribution function as defined in Section 3.5 represents the probability of finding another particle at a distance r away from a given particle; it is normalized so that g(r) approaches unity as $r \to \infty$.

The static structure factor S(k) is related to the radial distribution function via the Fourier transformation of Eq. (3.22). In Fig. 10.2 the values of S(k) as compiled by Galam and Hansen (1976) are plotted for various Γ.

For a weakly coupled OCP ($\Gamma \ll 1$), the random-phase approximation (RPA) employed in Appendix 5 is applicable, so that one finds Eq. (4.14). Departure of the static structure factor from its RPA values is measured by the local-field correction G(k), discussed earlier in Section 6.4 for a degenerate electron liquid. In the case of the classical OCP the local-field correction is given by

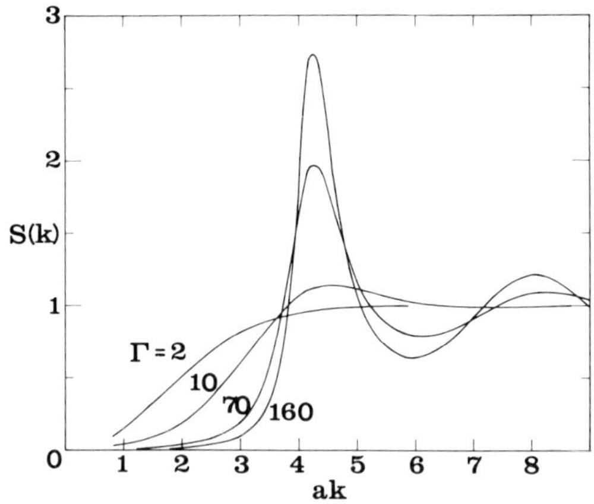


**Figure 10.2.** The static structure factor for OCP at various values of Γ.

$$G(k) = 1 + \left[1 - \frac{1}{S(k)}\right] \frac{k^2}{k_D^2} \quad . \tag{10.7}$$

The values of G(k) can be computed by substituting the Monte Carlo data of S(k) in Eq. (10.7); results are shown in Fig. 10.3.

The static dielectric function of a classical plasma is expressed, in reference to Eq. (6.30), as

$$\varepsilon(k, 0) = 1 + \frac{k_D^2}{k^2 - k_D^2 G(k)} \quad . \tag{10.8}$$

The long-wavelength behavior of the local-field correction is related to the isothermal compressibility of the OCP [e.g., Pines and Nozières 1966; cf. Eqs. (6.34) - (6.36) via

$$\lim_{k \to 0} \left(\frac{k_D}{k}\right)^2 G(k) = 1 - \frac{1}{k_B T} \left(\frac{\partial P}{\partial n}\right)_T \tag{10.9}$$

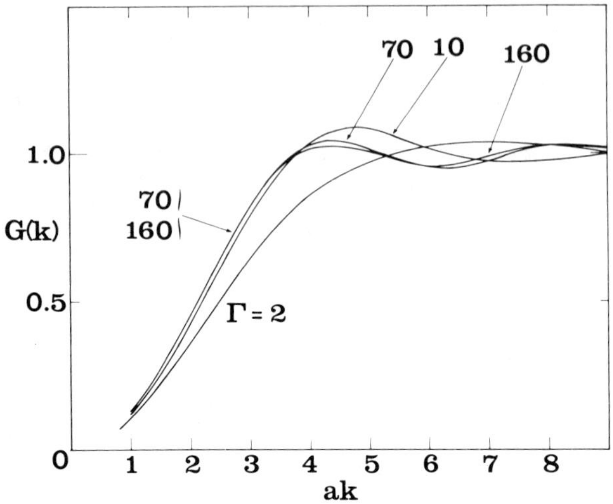


**Figure 10.3.** The local-field correction for OCP at various values of Γ.

where P is the pressure.

With the knowledge of the radial distribution function or the static structure factor, the excess internal energy density $U_{ex}$ of the OCP is calculated according to [cf. Eq. (6.6)]

$$\frac{U_{ex}}{nk_BT} = \frac{n}{2k_BT} \int d\vec{r} \, \frac{(Ze)^2}{r} [g(r) - 1]$$

$$= \frac{1}{4\pi^2 k_B T} \int d\vec{k} \, \frac{(Ze)^2}{k^2} [S(k) - 1] \quad . \tag{10.10}$$

Since the total internal energy density U of the system is given by

$$U = \frac{3}{2} nk_BT + U_{ex} \tag{10.11}$$

and the pressure by

$$P = nk_BT + U_{ex}/3 \quad , \tag{10.12}$$

various thermodynamic functions of the OCP can be determined once Eq. (10.10) is evaluated as a function of $\Gamma$. A recent result for U in the fluid phase has been given by Eq. (1.12).

The Helmholtz free energy F of the plasma can be calculated by integration with respect to the Coulomb coupling constant as

$$f(\Gamma) \equiv \frac{F}{nk_BT}$$

$$= \int_{\Gamma_1}^{\Gamma} \frac{U_{ex}}{nk_BT}\frac{d\Gamma'}{\Gamma'} + f(\Gamma_1) \quad . \tag{10.13}$$

The value of $\Gamma_1$ is to be chosen in the domain of the coupling constant where $f(\Gamma_1)$ is accurately known by some other means a priori. Equation (10.13) is the classical counterpart to the quantum version as expressed in Eq. (6.42).

Slattery, Doolen, and DeWitt (1982) thus determined the normalized free energy in the fluid phase as

$$\begin{aligned} f_F(\Gamma) &= -\,0.897744\Gamma + 3.80172\Gamma^{1/4} \\ &\quad -\,0.75824\Gamma^{-1/4} + 2.18513\ln\Gamma \\ &\quad -\,3.297336 \quad , \end{aligned} \tag{10.14}$$

and that in the body-centered-cubic (bcc) crystalline phase as

$$\begin{aligned} f_C(\Gamma) &= -\,0.895929\Gamma + 4.5\ln\Gamma - 1.8856 \\ &\quad -\,3225/2\Gamma^2 \quad . \end{aligned} \tag{10.15}$$

The intersection of Eqs. (10.14) and (10.15) gives the $\Gamma$ value at which the fluid-crystal transition or the **Wigner crystallization** takes places. This value was found to be

$$\Gamma_m = 178 \pm 1 \quad . \tag{10.16}$$

Since the radial distribution function g(r) describes the spatial correlation between two particles, the potential of mean force w(r) may be defined through the equation

$$g(r) = \exp[- w(r)/k_B T] \quad . \tag{10.17}$$

In a weakly coupled plasma it is sufficient to take the bare Coulomb interaction, $w(r) = (Ze)^2/r$. In a strongly coupled plasma, however, w(r) deviates significantly away from $(Ze)^2/r$ due to the effects of many-particle correlations. The **screening potential**,

$$H(r) = [(Ze)^2/r] - w(r) \quad , \tag{10.18}$$

measures the extent to which the effective potential w(r) between two particles is lowered from the bare Coulomb value $(Ze)^2/r$.

By substituting in Eq. (10.17) the values of g(r) such as shown in Fig. 10.1 and with the aid of Eq. (10.18), we may compute the screening potential H(r) evaluated on the basis of the Monte Carlo experiments. The result of such a computation has revealed an interesting feature. Figure 10.4 plots the values of $H(r)/[(Ze)^2/a]$ as a function of r/a for various Γ. Here we clearly observe that the data follow a straight line in the interval, $0.4 < r/a < 1.8$. This straight line may be expressed as

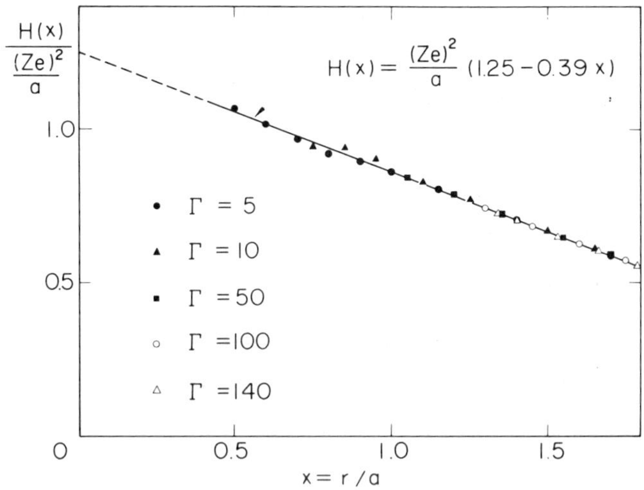


**Figure 10.4.** Linearity of the screening potential derived from the Monte Carlo simulation data.

$$H(r) = \frac{(Ze)^2}{a}\left[c_0 - c_1\left(\frac{r}{a}\right)\right] \tag{10.19}$$

for $4 \leq \Gamma \leq 160$. The choice of the coefficients, $c_0 = 1.25$ and $c_1 = 0.39$, reproduces the Monte Carlo data of Hansen (1973) within errors less than 4% (Itoh, Totsuji, and Ichimaru 1977). The coefficients are then mutually connected via

$$c_1 = (c_0/2)^2 \quad . \tag{10.20}$$

The linear relationship (10.19) and the connecting formula (10.20) were discovered by DeWitt, Graboske, and Cooper (1973) through the analysis of the Monte Carlo data of Brush et al. (1966), in connection with the study of the enhancement of thermonuclear reaction rate in dense plasmas, to be discussed later in Section 10.4. Such relations have been found to apply not only in the OCP but also in the binary mixtures of ions (Itoh, Totsuji, Ichimaru, and DeWitt 1979) and in the two-dimensional electron layers (Itoh, Ichimaru, and Nagano 1978; Totsuji 1978). Those features are in fact directly related to the existence of a latticelike short-range order in the fluid phase of a strongly coupled plasma (e.g., Ichimaru 1982).

### 10.3 Analytic Theories

In a theoretical treatment of interparticle correlations in a many-particle system, it is useful to start with the Bogoliubov-Born-Green-Kirkwood-Yvon (BBGKY) hierarchy of the kinetic equations (e.g., Ichimaru 1973; Hansen and McDonald 1976):

$$\left\{\frac{\partial}{\partial t} + \sum_{i=1}^{n} \left[\frac{1}{m}\vec{\mathbf{p}}_i \cdot \frac{\partial}{\partial \vec{\mathbf{r}}_i} + \left(\vec{\mathbf{F}}_i + \sum_{j(\neq i)}^{n} \vec{\mathbf{F}}_{ij}\right) \cdot \frac{\partial}{\partial \vec{\mathbf{p}}_i}\right]\right\} f^{(n)}(\vec{\mathbf{r}}^n, \vec{\mathbf{p}}^n, t)$$

$$= - \sum_{i=1}^{n} \int d\vec{\mathbf{r}}_{n+1} \int d\vec{\mathbf{p}}_{n+1} \vec{\mathbf{F}}_{i,n+1} \cdot \frac{\partial}{\partial \vec{\mathbf{p}}_i} f^{(n+1)}(\vec{\mathbf{r}}^{n+1}, \vec{\mathbf{p}}^{n+1}, t) \quad . \tag{10.21}$$

Here $f^{(n)}(\vec{\mathbf{r}}^n, \vec{\mathbf{p}}^n, t)$ is the n-particle distribution function in the 6n-dimensional phase space, $\vec{\mathbf{r}}^n = (\vec{\mathbf{r}}_1 \ldots \vec{\mathbf{r}}_n)$ and $\vec{\mathbf{p}}^n = (\vec{\mathbf{p}}_1 \ldots \vec{\mathbf{p}}_n)$; $\vec{\mathbf{F}}_i$ is any external force acting on particle i; and $\vec{\mathbf{F}}_{ij}$ is the force on i due to j.

Analytic theories accounting for the static correlations in a strongly coupled plasma have been concerned, to a certain degree, with a derivation of a relevant expression for the local-field correction G(k), as given by Eq. (10.7). This function measures the extent to which the particle interactions affect the static correlational properties in such a plasma.

In the polarization potential model of condensed matter, explained in Section 6.4, the local-field correction enters the expression for the scalar polarization potential; characterizing that portion of the restoring forces responsible for the collisionless part of the excitation spectrum which couples directly to the density fluctuations in the system.

Singwi, Tosi, Land, and Sjölander (STLS) (1968) advanced a detailed theory of electron correlations at metallic densities, in which a semiclassical calculation of G(k) was given as a functional of S(k). These authors truncated the BBGKY hierarchy of the kinetic equations Eq. (10.21) by assuming that the two-particle distribution function, was expressed in terms of the single-particle distribution function and the equilibrium, radial distribution function. Solving a linear density-response problem in the first equation of the BBGKY hierarchy, they derived Eq. (6.39) for the expression of the local-field correction.

A combination of Eqs. (6.39) and (10.7) then makes a self-consistent integral equation for determination of S(k) or G(k) in the classical OCP. A numerical solution to this equation was carried out by Berggren (1970). When the STLS theory was applied to a weakly coupled OCP ($\Gamma \ll 1$), however, it failed to reproduce the known $\Gamma$ expansion Eq. (1.11) of the internal energy beyond the Debye-Hückel term, i.e., the second term on the right-hand side of Eq. (1.11).

The local-field correction can also be obtained directly from a solution to the second equation of the BBGKY hierarchy. The Born-Green equation (e.g., Feynman 1972) relating between the pair and ternary correlation functions in thermodynamic equilibrium can be transformed in a form (Tago, Utsumi, and Ichimaru 1981)

$$G(k) = -\frac{1}{nS(k)} \int \frac{d\vec{q}}{(2\pi)^3} K(\vec{k}, \vec{q}) [S(|\vec{k} - \vec{q}|)$$

$$- 1 + \frac{1}{2} \tilde{h}^{(3)}(\vec{k} - \vec{q}, \vec{q})] \quad . \qquad (10.22)$$

Here

$$K(\vec{k}, \vec{q}) = \frac{\vec{k} \cdot \vec{q}}{q^2} + \frac{\vec{k} \cdot (\vec{k} - \vec{q})}{|\vec{k} - \vec{q}|^2} \quad , \qquad (10.23)$$

$$\tilde{h}^{(3)}(\vec{k}_1, \vec{k}_2) = n^2 \int d\vec{r}_{13} \int d\vec{r}_{23} h^{(3)}(r_{12}, r_{23}, r_{31})$$

$$\times \exp(-i\vec{k}_1 \cdot \vec{r}_{13} - i\vec{k}_2 \cdot \vec{r}_{23}) \quad ,$$

$$(10.24)$$

$$\vec{r}_{ij} = \vec{r}_i - \vec{r}_j \quad ,$$

and $h^{(3)}(r_{12}, r_{23}, r_{31})$ refers to the radial part of the ternary correlation function (e.g., Ichimaru 1973). In terms of the ternary radial-distribution function $g^{(3)}(r_{12}, r_{23}, r_{31})$ and the radial part of the pair correlation function

$$h(r) = g(r) - 1 \quad , \qquad (10.25)$$

it is expressed as

$$h^{(3)}(r_{12}, r_{23}, r_{31}) = g^{(3)}(r_{12}, r_{23}, r_{31}) - 1$$

$$- h(r_{12}) - h(r_{23}) - h(r_{31}) \quad . \qquad (10.26)$$

The hierarchy can be terminated if the ternary correlation function is expressed in terms of the pair correlation function in Eq. (10.22); a self-consistent integral equation will result through a combination with Eq. (10.7).

Ichimaru (1970) introduced the convolution approximation in Eq. (10.22), so that the ternary correlation function is expressed as

$$h_{CA}^{(3)}(r_{12}, r_{23}, r_{31}) = h(r_{12})h(r_{23}) + h(r_{23})h(r_{31}) + h(r_{31})h(r_{12}) + n\int d\vec{r}_4 h(r_{24})h(r_{34}) \quad . \tag{10.27}$$

Graphically Eq. (10.27) may be described as in Fig. 10.5. It was shown by O'Neil and Rostoker (1965) that the BBGKY long-range solution of the ternary correlation function for an OCP in the weak-coupling limit ($\Gamma \ll 1$) takes the form Eq. (10.27), where the pair correlation function is given by the Debye-Hückel formula,

$$h(r) = -\frac{(Ze)^2}{k_B T r} \exp\left(-\frac{r}{\lambda_D}\right) \quad . \tag{10.28}$$

By adopting the convolution approximation (10.27) in the second BBGKY solution (10.22), one thus automatically guarantees accuracy of the resulting long-range solution of the pair correlation function to the first two terms in the $\Gamma$ expan sion. Substitution of Eq. (10.27) in Eq. (10.22) yields Eq.

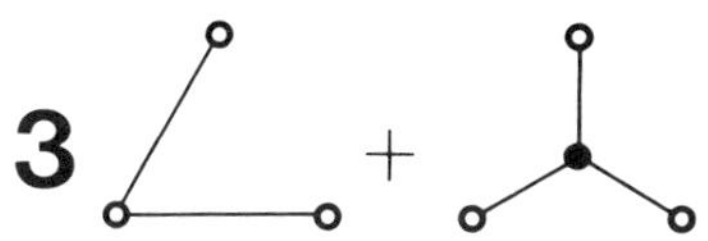


**Figure 10.5.** Convolution approximation for the ternary correlation function.

(6.40), where salient features in its numerical solution were also described.

For the classical OCP, the hypernetted chain (HNC) equation developed in the theory of liquids (e.g., van Leeuwen, Groeneveld, and De Boer 1959; Morita 1960) is known to provide a theoretical description much more accurate than the Percus-Yevick equation (Percus and Yevick 1958) or the integral equations discussed earlier in this section. Let us therefore look at this HNC scheme as applied to the strongly coupled OCP problem.

In terms of the Mayer cluster expansion (e.g., Rice and Gray 1965; Hansen and McDonald 1976), the radial distribution function is exactly expressed as

$$g(r) = \exp\{-[\phi(r)/k_B T] + N(r) + B(r)\} \quad . \qquad (10.29)$$

Here $\phi(r)$ represents the interparticle potential [for the OCP, $\phi(r) = (Ze)^2/r$]; $N(r)$ refers to the sum of the nodal diagrams; and $B(r)$, that of the bridge diagrams. The nodal function $N(r)$ can be written as

$$N(r) = h(r) - c(r) \quad , \qquad (10.30)$$

where $c(r)$ is the direct correlation function; this function is connected to the pair correlation function $h(r)$ via the Ornstein-Zernike relation,

$$h(r) = c(r) + n \int d\vec{r}' c(|\vec{r} - \vec{r}'|)h(r') \quad . \qquad (10.31)$$

The HNC approximation corresponds to assuming $B(r) = 0$ in Eq. (10.29). One then has a closed set of equations, so that $h(r)$ and $c(r)$ may be calculated from a solution of Eqs. (10.25), (10.31), and the HNC equation

$$g(r) = \exp\{-[\phi(r)/k_BT] + h(r) - c(r)\} \quad . \tag{10.32}$$

In the HNC approximation the local-field correction takes the form

$$G(k) = -\frac{1}{n}\int \frac{d\vec{q}}{(2\pi)^3}\,\frac{\vec{k}\cdot\vec{q}}{q^2}\,[S(|\vec{k}-\vec{q}|) - 1] \times \{1 + [1 - G(q)][S(q) - 1]\} \quad . \tag{10.33}$$

We note here that the STLS scheme Eq. (6.39) may be obtained by setting $G(q) = 1$ on the right-hand side of Eq. (10.33); the convolution formula, Eq. (6.40), can be derived by assuming $G(q) = 0$ there.

The numerical solution to the HNC equation for the OCP was carried out by Springer, Pokrant, and Stevens (1973) for the radial distribution function and the internal energy in the domain $0.05 \le \Gamma \le 50$; this solution was subsequently extended more accurately by Ng (1974) in the domain $20 \le \Gamma \le 7000$. Those HNC data for the excess internal energy were fitted by DeWitt (1976) with the result

$$\frac{U}{nk_BT} = -0.900470\Gamma + 0.2688263\Gamma^{1/2} + 0.07199925\ln\Gamma + 1.5537919 \tag{10.34}$$

for $\Gamma > 1$.

Quantitatively the HNC internal energy agrees with the Monte Carlo data over the entire fluid range $\Gamma < \Gamma_m$ with discrepancies of less than 1%. The main reason for the apparent good agreement of the HNC and Monte Carlo results is that the static Madelung-like term in the HNC energy, i.e., the first term on the right-hand side of Eq. (10.34), is very close to that for the Monte Carlo fluid energy in Eq. (1.12). The static term from HNC is nearly identical to the ion-sphere result ($-0.9\,\Gamma$) --cf. Section 2.4--and is 0.5% below the bcc lattice value ($-0.895929\Gamma$) in Eq. (10.15), while the static term in Eq. (1.12) is 0.2% below the bcc lattice value. The remaining energy, to be called the thermal-energy function, from HNC is significantly different from the Monte Carlo form, $\Gamma^{1/2}$ vs. $\Gamma^{1/4}$, and the HNC thermal-energy function is numerically larger substantially than the Monte Carlo thermal-energy function.

The HNC approximation, though superior to many of the analytic schemes proposed for the treatment of the OCP, still exhibits a systematic departure from the Monte Carlo results in the strong-coupling domain $\Gamma > 1$, as we have seen in the comparison of the thermal-energy function. As we observe in Figs. 10.6 and 10.7, the behaviors of the HNC radial distribution function and screening potential differ somewhat from those indicated in the Monte Carlo simulation; the amplitudes of the oscillations in $g(r)$ are usually underestimated in HNC. A further improvement over the HNC scheme is therefore a meaningful project in the analytic theory of the OCP.

Obviously from the derivation of the HNC equation (10.32), a possible area of improvement lies in the treatment of the bridge function $B(r)$. Two frontal developments have been advanced recently in this area: One is based on the use of parametrized hard-sphere bridge functions (Rosenfeld and Ashcroft 1979), and the other relies on an approximate evaluation of the bridge function for the OCP in the framework of the density-functional formalism (Iyetomi and Ichimaru 1983b; Iyetomi 1984). As the comparison in Figs. 10.6 and 10.7 illustrates, both theories predict the thermodynamic quantities and the radial distribution functions nearly indistinguishable

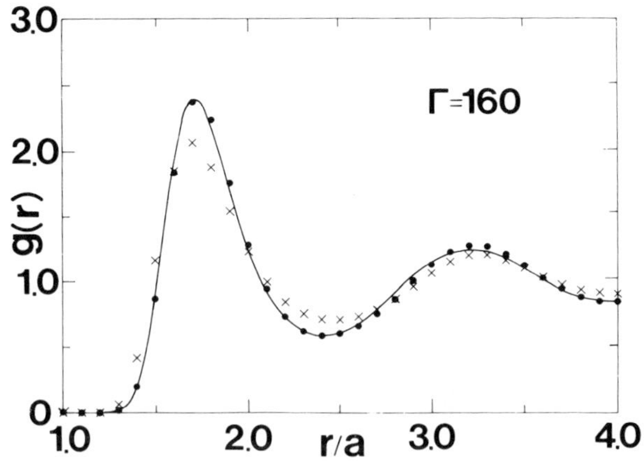


**Figure 10.6.** The radial distribution for OCP at $\Gamma = 160$. The filled circles represent the Monte Carlo results due to Slattery, Doolen, and DeWitt (1980); the crosses, the HNC results. The solid curve depicts an improved HNC calculation based on the scheme of Iyetomi and Ichimaru (1983b).

from the exact Monte Carlo results (e.g., Ichimaru 1982).

## 10.4 Enhancement of Thermonuclear Reaction Rate

Enhancement of thermonuclear reaction rate arising from Coulomb correlations in strongly coupled plasmas has important consequences in various aspects of stellar evolution such as carbon ignition in degenerate cores. The screening potential defined in Eq. (10.18) acts to lower the repulsive potential barrier between two reacting nuclei (Salpeter and Van Horn 1969).

The penetration probability p(v) of the Coulomb barrier for the reacting nuclei with the relative velocity v and the reduced mass M/2 may be given by the WKB integral (Van Horn and Salpeter 1967) as

$$p(v) = \exp\left\{-\frac{2\sqrt{M}}{\hbar}\int_0^r dr'\left[w(r') - \frac{Mv^2}{4}\right]^{1/2}\right\} , \qquad (10.35)$$

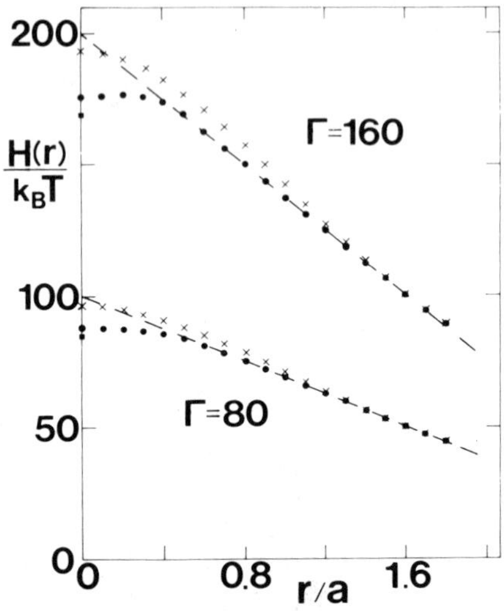


**Figure 10.7.** The screening potential for OCP at $\Gamma = 80$ and 160. The dashed lines refer to the Monte Carlo fit Eq. (10.19) and its linear extrapolation. Crosses are the HNC values; the filled circles, the improved HNC calculation of Iyetomi and Ichimaru (1983b). Squares represent the ion-sphere result (Salpeter 1954), $H(0)/k_BT = 1.057\Gamma$.

where $w(r')$ is the potential of mean force [cf. Eq. (10.17)] and $r$ is the classical turning radius determined from

$$w(r) = Mv^2/4 \quad . \tag{10.36}$$

Separating the strongly v-dependent part, we may write the nuclear reaction rate R as

$$R = \int_0^\infty dv\ F(v)\exp(-\ Q) \quad , \tag{10.37}$$

where $\exp(-\ Q)$ is the product between $p(v)$ and the Boltzmann factor $\exp(-\ Mv^2/4k_BT)$; $F(v)$ is the part weakly dependent on v.

Because of the strong v dependence of exp(- Q), the integral Eq. (10.37) may be evaluated through expansion of Q around its minimum value $Q(v_0)$ at $v_0$, where $dQ(v_0)/dv_0 = 0$ (the Gamow peak); we thus have

$$R \simeq F(v_0)\int_{-\infty}^{\infty} dv \exp\{ - Q(v_0) - \frac{1}{2}(v - v_0)^2 \frac{d^2}{dv_0^2} Q(v_0)\} \quad . \tag{10.38}$$

Since the relative velocity v is connected with the classical turning radius r via Eq. (10.36), one finds it convenient to regard Q as a function of r; i.e.,

$$Q(r) = \frac{w(r)}{k_B T} + \frac{2\sqrt{M}}{\hbar}\int_0^r dr' [w(r') - w(r)]^{1/2} \quad , \tag{10.39}$$

and thereby to carry out a statistical average with respect to the distribution of r. The latter distribution can be evaluated by substituting Eqs. (10.18) and (10.19) in Eq. (10.17). Introducing a normalized distance

$$\rho = r/[\frac{\hbar^2}{M(Ze)^2} (\frac{2\tau}{3\pi})^2] \quad , \tag{10.40}$$

and the dimensionless temperature parameter

$$\tau = [(\frac{27\pi^2}{4})\frac{M(Ze)^4}{\hbar^2 k_B T}]^{1/3} \quad , \tag{10.41}$$

one expresses the result to a good accuracy as (Itoh, Totsuji and Ichimaru 1977)

$$Q(\rho) \simeq \frac{\tau}{3}[\frac{1}{\rho} + 2\rho^{1/2} - \frac{3\Gamma}{\tau} c_0 + (\frac{3\Gamma}{\tau})^2 c_1\rho - \frac{1}{4}(\frac{3\Gamma}{\tau})^2 c_1 \rho^{5/2}] \quad . \tag{10.42}$$

The enhancement factor is the ratio of Eq. (10.38) calculated with the screening potential to that without it [i.e., $w(r) = (Ze)^2/r$]. The enhancement factor due to strong screening is thus given by $\exp[\tau - Q(\rho_0)]$, where $\rho_0$ is the value of $\rho$ at which $Q(\rho)$ in Eq. (10.42) is minimized; $(3\Gamma/\tau)\rho_0 a$ then represents approximately the classical turning radius at the Gamow peak. The values of $\tau - Q(\rho_0)$ so computed can be reproduced by the following formula within errors less than 1%:

$$\tau - Q(\rho_0) = 1.25\Gamma - 0.095\tau(3\Gamma/\tau)^2 \quad . \tag{10.43}$$

The effects of possible deviation of $H(r)$ from Eq. (10.19) in the vicinity of $r = 0$ (cf. Fig. 10.7) have also been investigated (Alastuey and Jancovici 1978; Itoh et al., 1979) and found to induce a negligible correction to Eq. (10.43). For a carbon plasma ($Z = 6$) at $T = 10^8$ K with mass density $10^9$ g/cm$^3$, the resulting enhancement factor turns out to be $\sim 6 \times 10^{16}$; at $T = 5 \times 10^7$ K with the same mass density it becomes $\sim 3 \times 10^{30}$.

### 10.5 Electric Resistivity

In a high-density plasma the frequency of Coulomb scattering between charged particles, considered in Chapter 2, may be drastically influenced by the presence of strong interionic correlations. Correspondingly the transport coefficients such as the electric resistivity may exhibit dependence on the plasma parameters quite different from those appropriate to a weakly coupled plasma. In this section, to illustrate these features, we take up the problem of calculating the electric resistivity in plasmas.

We begin with the kinetic equation (A3.11) for the particles of "σ" species. We assume only the electrostatic Coulomb interactions between the charged particles, so that the Fourier components of the electromagnetic fluctuations on the right-hand side of Eq. (A3.11) are given by

$$i\vec{k} \cdot \delta\vec{E}(\vec{k}, \omega) = \Sigma_\sigma 4\pi e_\sigma \delta n_\sigma(\vec{k}, \omega) \quad ,$$

$$\delta\vec{B}(\vec{k}, \omega) = 0 \quad . \tag{10.44}$$

For the calculation of the dc electric resistivity we may assume that $\partial/\partial t = 0$ and $\partial/\partial\vec{r} = 0$ on the left-hand side of Eq. (A3.11). Multiplying both sides of Eq. (A3.11) by $\vec{p}$ and integrating the resulting equation over $\vec{p}$, we find after partial integrations

$$\langle\vec{E}\rangle + \frac{1}{c}\,\vec{u}_\sigma \times \langle\vec{B}\rangle = \frac{1}{n_\sigma}\sum_{\sigma'} \int \frac{d\vec{k}}{(2\pi)^3} \int d\omega\, \frac{4\pi e_{\sigma'}\vec{k}}{k^2}\, \mathrm{Im}\, S_{\sigma\sigma'}(\vec{k}, \omega) \quad . \tag{10.45}$$

Here

$$\vec{u}_\sigma = \frac{1}{m_\sigma n_\sigma} \int d\vec{p}\;\vec{p}\;F_\sigma(X;\, t) \tag{10.46}$$

is the flow velocity of the particles of σ species, and

$$S_{\sigma\sigma'}(\vec{k}, \omega) = \frac{1}{2\pi} \int d\vec{r} \int_{-\infty}^{\infty} dt \langle \delta n_\sigma(\vec{r}' + \vec{r}, t' + t)\,\delta n_{\sigma'}(\vec{r}', t')\rangle \times \exp(-i\vec{k} \cdot \vec{r} + i\omega t) \tag{10.47}$$

refers to the Fourier transformation of the density-density correlation function between the $\sigma$ and $\sigma'$ species (cf. Appendix 8).

We define the electric resistivity $\eta_\sigma$ associated with the flow of the $\sigma$ species by setting the right-hand side of Eq. (10.45) equal to $\eta_\sigma e_\sigma n_\sigma \vec{u}_\sigma$; we thus obtain

$$\eta_\sigma = \sum_{\sigma'} \frac{e_{\sigma'}}{e_\sigma} \int \frac{d\vec{k}}{(2\pi)^3} \int d\omega \, \frac{4\pi \vec{k} \cdot \vec{u}_\sigma}{(n_\sigma u_\sigma k)^2} \, \mathrm{Im}\, S_{\sigma\sigma'}(\vec{k}, \omega) \quad . \qquad (10.48)$$

When the electrons have a net drift velocity $\vec{u}$ relative to the ions in an electron-ion two-component plasma, we find the electric resistivity expressed as

$$\eta_e = - \frac{4\pi Z}{(n_e u)^2} \int \frac{d\vec{k}}{(2\pi)^3} \int d\omega \, \frac{\vec{k} \cdot \vec{u}}{k^2} \, \mathrm{Im}\, S_{ei}(\vec{k}, \omega) \quad . \qquad (10.49)$$

These formulas are to be evaluated in the limit of $u \to 0$.

To write the electron-ion correlation $S_{ei}(\vec{k}, \omega)$ in terms of the ion-ion correlations $S_{ii}(\vec{k}, \omega)$, we evoke the idea of the polarization-potential approximation as considered in Section 6.4. Regarding the ion-density fluctuations $\delta n_i(\vec{k}, \omega)$ as the external perturbation to the electron system, that is

$$(\phi_{ext})_{\vec{k}\omega} = (4\pi Ze/k^2)\,\delta n_i(\vec{k}, \omega) \quad , \qquad (10.50)$$

we substitute Eq. (10.50) in Eq. (6.27), to find the induced electron-density fluctuations $\delta n_e(\vec{k}, \omega)$ given by

$$\delta n_e(\vec{k}, \omega) = \frac{v_{ei}(k)\chi_0(\vec{k}, \tilde{\omega})}{\tilde{\varepsilon}(\vec{k}, \tilde{\omega})} \delta n_i(\vec{k}, \omega) \quad . \tag{10.51}$$

Here

$$v_{ei}(k) = - 4\pi Ze^2/k^2 \quad , \tag{10.52}$$

$\chi_0(\vec{k}, \tilde{\omega})$ is the free-electron polarizability with the displaced frequency

$$\tilde{\omega} = \omega - \vec{k} \cdot \vec{u} \quad , \tag{10.53}$$

and

$$\tilde{\varepsilon}(\vec{k}, \tilde{\omega}) = 1 - (4\pi e^2/k^2)[1 - G(\vec{k})]\chi_0(\vec{k}, \tilde{\omega}) \tag{10.54}$$

is the electronic screening function of the ionic potential (e.g., Heine, Nozières, and Wilkins 1966; Hedin and Lundqvist 1971; Kukkonen and Overhauser 1979; Ichimaru and Iyetomi 1983). In the derivation of Eq. (10.51), the approximations leading to Eq. (6.30) have been adopted, so that G(k) in Eq. (10.54) represents the local-field correction of the electrons.

With the aid of Eq. (10.51) we may express

$$S_{ei}(\vec{k}, \omega) = v_{ei}(k) \frac{\chi_0(\vec{k}, \tilde{\omega})}{\tilde{\varepsilon}(\vec{k}, \tilde{\omega})} S_{ii}(\vec{k}, \omega) \quad . \tag{10.55}$$

Since $S_{ii}(\vec{k}, \omega)$ is a real function, substitution of Eq. (10.55) in Eq. (10.49) yields

$$\eta_e = \frac{1}{(en_e u)^2} \int \frac{d\vec{k}}{(2\pi)^3} \int d\omega\, \vec{k} \cdot \vec{u} \left| \frac{v_{ei}(k)}{\tilde{\varepsilon}(\vec{k}, \tilde{\omega})} \right|^2 S_{ii}(\vec{k}, \omega) \, \mathrm{Im}\, \chi_0(\vec{k}, \tilde{\omega}) . \tag{10.56}$$

From the point of view of the dynamic behaviors of the electrons, represented by $\tilde{\varepsilon}(\vec{k}, \tilde{\omega})$ and $\chi_0(\vec{k}, \tilde{\omega})$, the ions may be regarded as slowly moving and the frequency spectrum $S_{ii}(\vec{k}, \omega)$ of their density fluctuations as confined almost in the vicinity of $\omega = 0$. The $\omega$ integration in Eq. (10.56) will be carried out with this feature in mind.

When the electrons are completely degenerate, Eqs. (6.17) and (6.18) apply for the calculation of $\chi_0(\vec{k}, \omega)$; we thus have

$$\mathrm{Im}\, \chi_0(\vec{k}, \omega) = -\frac{m^2}{2\pi\hbar^3} \frac{\omega}{k} \qquad \left(0 \le \omega \le kv_F - \frac{\hbar k^2}{2m}\right) . \tag{10.57}$$

Substitution of Eq. (10.57) in Eq. (10.56) leads to the **Ziman formula** (Ziman 1961) for electric resistivity,

$$\eta_e = \frac{m^2 n_i}{12\pi^3 \hbar^3 e^2 n_e^{\,2}} \int_0^{2k_F} dk\, k^3 \left| \frac{v_{ei}(k)}{\tilde{\varepsilon}(\vec{k}, 0)} \right|^2 S_{ii}(\vec{k}) , \tag{10.58}$$

where

$$S_{ii}(\vec{k}) = \frac{1}{n_i} \int d\omega\, S_{ii}(\vec{k}, \omega) \tag{10.59}$$

is the static structure factor of the ions. Several investigators in the past (e.g., Hubbard and Lampe 1969; Stevenson and Ashcroft 1974; Flowers and Itoh 1976; Minoo, Deutsch, and

Hansen 1976; Ichimaru and Iyetomi 1983; Itoh, Mitake, Iyetomi, and Ichimaru 1983) computed the electric resistivity of dense matter on the basis of the Ziman formula, with various degrees of approximations to the evaluations of $\tilde{\varepsilon}(\vec{k}, 0)$ and $S_{ii}(\vec{k})$. Figure 10.8 shows the results of such a calculation (Iyetomi and Ichimaru 1983a) for $^{56}Fe$ plasmas at $T = 10^7$ K appropriate to the outer crustal matter of a neutron star.

When the electrons assume a nondegenerate Maxwellian distribution, Eq. (A5.23) is applicable, and we have

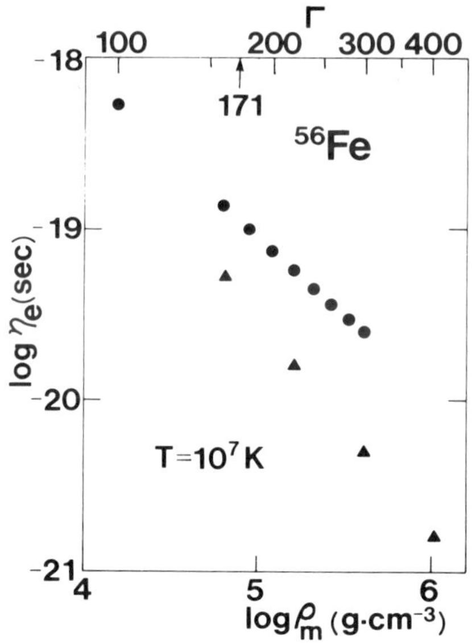


**Figure 10.8.** The electric resistivity $\eta_e$ of $^{56}Fe$ plasma at $T = 10^7$ K as a function of the mass density $\rho_m$ or the Coulomb coupling constant $\Gamma$. Filled circles refer to the calculation in the homogeneous fluid state; the triangles, in the crystalline state (Flowers and Itoh 1973). After Iyetomi and Ichimaru (1983a).

$$\mathrm{Im}\ \chi_0(\vec{k}, \omega) = -\left(\frac{\pi}{2}\right)^{1/2} \frac{n_e}{k_B T} \frac{\omega}{k(k_B T/m)^{1/2}} \exp\left[-\frac{\omega^2}{2k^2(k_B T/m)}\right] \ . \tag{10.60}$$

For those classical-electron cases, we thus find

$$\eta_e = \left(\frac{m}{2\pi k_B T}\right)^{3/2} \frac{n_i}{3e^2 m n_e} \int_0^{k_{max}} dk\ k^3 \left|\frac{v_{ei}(k)}{\tilde{\varepsilon}(\vec{k}, 0)}\right|^2 S_{ii}(\vec{k}) \ . \tag{10.61}$$

In this expression we had to introduce an upper limit $k_{max}$ for the k integration, to avoid a logarithmic divergence. When the possibility of different electron and ion temperatures is permitted for plasmas with Z = 1, the calculation of Eq. (10.61) yields (Tange and Ichimaru 1974)

$$\eta_e = \frac{4(2\pi)^{1/2} e^2 m^{1/2}}{3(k_B T_e)^{3/2}} \left[2\ln\Lambda - \left(1 + \frac{T_i}{T_e}\right)\ln\left(1 + \frac{T_e}{T_i}\right) + \ldots\right] \tag{10.62}$$

where the Coulomb logarithm is identified as

$$\ln\Lambda = \frac{1}{2}\left(1 + \frac{T_i}{T_e}\right)\ln\left(\frac{k_{max}^2}{k_{De}^2}\right) \ . \tag{10.63}$$

The magnitude of the electric resistivity calculated according to the classical scheme of Eq. (10.61) is known (e.g., Kaneko 1960; Williams and DeWitt 1969; Tange and Ichimaru 1974) to take on a magnitude 1.97 times as large as the Spitzer value (Spitzer and Härm 1953; Spitzer 1962) for Z = 1, because possible deformation of the electronic distribution is not appropriately taken into account in the calculation leading to Eq. (10.56). When the electrons in the plasma under consideration obey the classical statistics, one must take account of such a factor in the evaluations of the electric resistivity and other transport coefficients (e.g., Ichimaru and Tanaka 1985).

The artificial introduction of the upper limit $k_{max}$ in Eq. (10.61) becomes unnecessary when we retain the quantum nature in the statistics of the electrons (Boercker, Rogers, and DeWitt 1982; Ichimaru and Tanaka 1985). Thus for arbitrary degeneracy of the electrons at nonzero temperatures, we set the Fermi distribution

$$f_0(q) = \frac{1}{\exp[(\varepsilon_q - \mu)/k_BT] + 1} \tag{10.64}$$

with $\varepsilon_q = \hbar^2q^2/2m$, where the chemical potential $\mu$ is determined from the normalization

$$\frac{n_e}{2} = \int_0^\infty dq \, \frac{q^2}{2\pi^2} f_0(q) \quad . \tag{10.65}$$

Equation (10.56) then takes a convergent expression

$$\eta_e = \frac{m^2n_i}{12\pi^3\hbar^3e^2n_e^2} \int_0^\infty dk \; k^3 f_0\left(\frac{k}{2}\right) \left|\frac{v_{ei}(k)}{\tilde{\varepsilon}(\vec{k}, 0)}\right|^2 S_{ii}(\vec{k}) \quad . \tag{10.66}$$

Explicit calculation of Eq. (10.66) has been carried out for 21 parametric combinations ranging $0.003 \le \Gamma \le 2$ and $\theta$ ($\equiv k_BT/E_F$) = 0.1, 1.0 and 10, with Z = 1 (Ichimaru and Tanaka 1985). Writing

$$\eta_e = \frac{4L}{3\sqrt{\pi}\,\theta^{3/2}} \eta_0 \tag{10.67}$$

with

$$\eta_0 \equiv \frac{4m^2e^2}{3\pi\hbar^3 n} = \frac{16}{9}\frac{\hbar^3}{me^4} r_s{}^3 = 38.6\ r_s{}^3 \quad [\mu\Omega \cdot \mathrm{cm}] \tag{10.68}$$

we may parametrize the generalized Coulomb logarithm L in an analytic formula applicable for $\Gamma \leq 2$ and $\theta \geq 0.1$:

$$L(\Gamma,\ \theta) = \frac{a(\theta)\ln\Gamma + b(\theta) + c(\theta)\Gamma}{1 + d(\theta)\Gamma^3} \tag{10.69}$$

where

$$a(\theta) = \frac{-\ \theta^{3/2}}{2\theta^{3/2} - 0.57923\theta + 0.23272\theta^{1/2} + 1.4853} \tag{10.70}$$

$$b(\theta) = \frac{\theta^3(\frac{1}{2}\ \ln\theta - 0.18603) + 1.2704\theta^{3/2}}{\theta^3 + 1.8993\theta^2 + 4.3243\theta + 1} \tag{10.71}$$

$$c(\theta) = \theta^{3/2}\ \frac{0.69460\theta^{1/2} + 0.24228}{\theta^2 + 1.7768} \tag{10.72}$$

$$d(\theta) = \theta^{3/2}\ \frac{0.13550\theta^{1/2} + 0.083521}{\theta^2 + 0.36797}\ . \tag{10.73}$$

The resultant expression (10.67) reproduces all the computed values of Eq. (10.66) with digressions of less than 2%.

# Chapter 11

# Free Electron Laser

When an electron has a relativistic high energy, it couples strongly with the electromagnetic radiation field. One of the reasons for strong coupling is the relative ease in satisfying the resonance conditions (2.40) as the electron velocity approaches light velocity. It may also be pointed out that the electromagnetic field in the frame of a high-speed electron differs remarkably from that in the laboratory frame owing to the relativistic effect. Consequently, the electromagnetic wave (a photon) scattered by a relativistic electron receives a substantial enhancement of its frequency (energy). This is the inverse Compton process, playing an important part in the interpretation of various astrophysical phenomena such as X-ray stars. This inverse Compton process is also positively utilized in the free-electron laser experiment. In this chapter we study those radiative processes of relativistic electrons, related to the free-electron laser experiment.

## 11.1 Stanford Experiments

We consider a helical coil (length L, pitch of the winding $\lambda_q$), and choose the z axis along its axis. Place another identical coil and rotate it by 180 degrees around the z axis with respect to the first coil. Connect this double-helix coil at one end and let a steady electric current flow through it. The induced magnetic field along the z axis is expressed as

$$\vec{\mathbf{B}}(z) = \delta B\{\hat{\mathbf{x}} \cos(qz) + \hat{\mathbf{y}} \sin(qz)\} \quad . \tag{11.1}$$

The z component of the field is canceled because of the current flowing in opposite directions in the double helix. Here

$$q = 2\pi/\lambda_q \tag{11.2}$$

and $\hat{\mathbf{x}}$ and $\hat{\mathbf{y}}$ are the unit vectors in the respective directions. A magnetic field in the shape of Eq. (11.1) is called the **wiggler.**

Experiments were performed in W. W. Hansen High Energy Physics Laboratory at Stanford (Elias, Fairbank, Madey, Schwettman, and Smith 1976), where a 24 MeV electron beam is injected along the z axis of a wiggler magnetic field with $L = 5.2$ m, $\lambda_q = 3.2$ cm, $\delta B = 2.4$ k gauss (see Fig. 11.1). Since the rest-mass energy of an electron is $mc^2 = 0.511$ MeV, this beam is in the relativistic high-energy domain. The electron density in the beam was not continuous but formed "bunches"; the peak value of the instantaneous current was 70 mA. Intensity of electromagnetic wave radiated into an opening angle (solid angle) $5 \times 10^{-6}$ sr in the z direction was measured by a detector with the central wavelength 10.6 μm and resolution of 0.2%. When the energy of the incident electron beam was varied by approximately 2% around 24 MeV, a spectrum as shown in Fig. 11.2 (a) was detected. The relative spread of the spectrum was 0.4% and a maximum output of $4 \times 10^{-6}$ W was observed at the

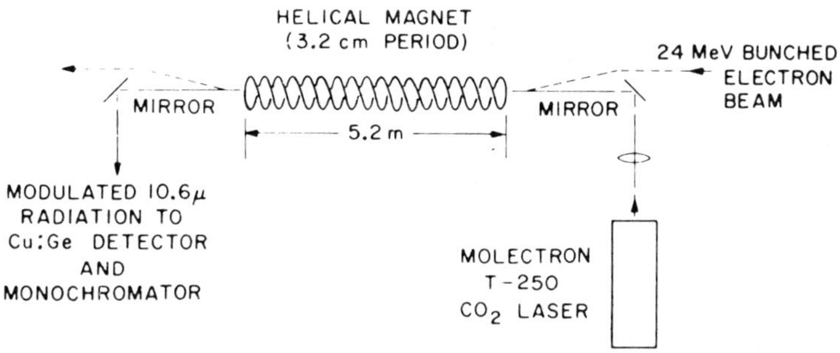


**Figure 11.1.** Interaction between an electron beam and a wiggler field--spontaneous emission and amplification of electromagnetic radiation. After Elias et al. (1976).

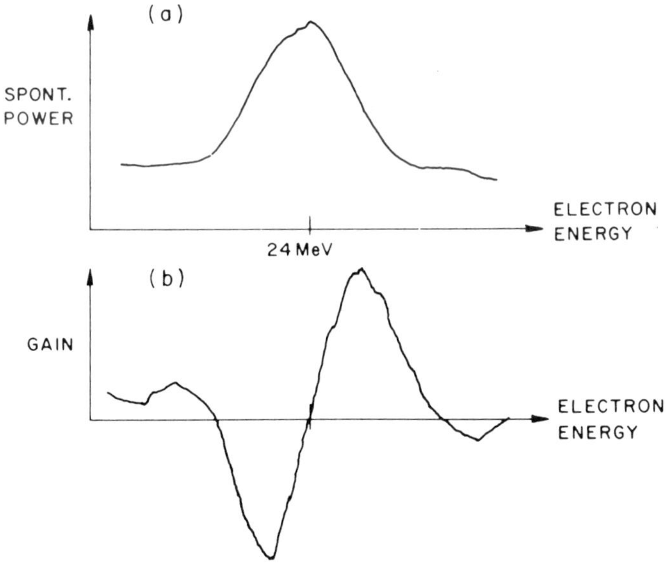


**Figure 11.2.** Rates of spontaneous emission and amplification at 10.6 μm as functions of the electron energy. After Elias et al. (1976).

electric current of 70 mA.

Next an electromagnetic wave with wavelength 10.6 μm produced by a $CO_2$ laser was injected along the z axis of this apparatus and the rate of amplification (or attenuation) was measured. Figure 11.2 (b) shows the measured result as a function of the electron energy. The rate of amplification for a passage of the electromagnetic wave over the total coil length reached a maximum value of 7% at the peak current value of 70 mA. These experiments clearly demonstrated that a relativistic electron beam injected into a wiggler field can produce a spontaneous emission of electromagnetic wave and can amplify an electromagnetic wave introduced from an external source.

About one year later, the first experimental demonstration of the **free electron laser** was successfully carried out at the Stanford Laboratory (Deacon, Elias, Madey, Ramian, Schwettman, and Smith 1977). As Fig. 11.3 shows, the wiggler field was the same as that in Fig. 11.1 both in strength and in shape. Energy of the electron beam and the peak current value were raised to 43.5 MeV and 2.6 A. The interaction domain between the beam and the wiggler field constituted a resonator of the electromagnetic wave with reflecting mirrors located at both ends.

Figure 11.4 (bottom) shows the spectral distribution of radiation

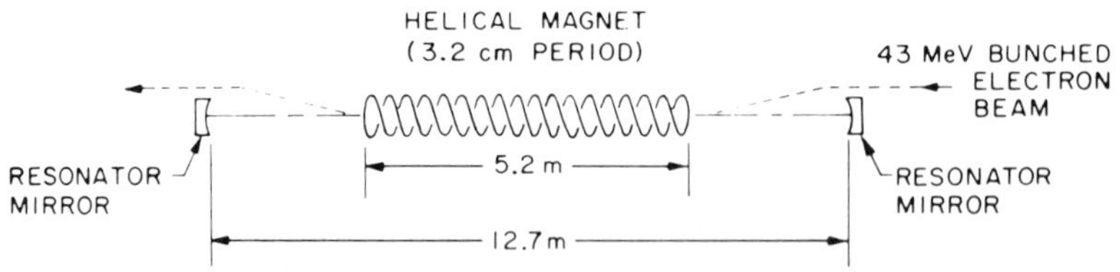


**Figure 11.3.** Schematic diagram of the free-electron laser experiment. After Deacon et al. (1977).

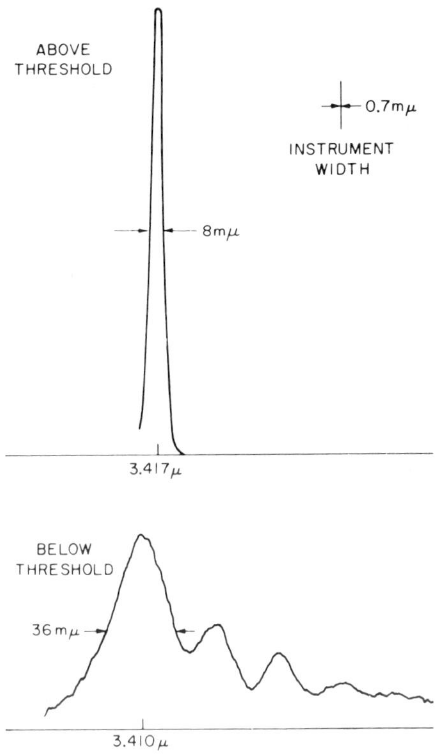


**Figure 11.4.** Emission spectrum of the laser oscillator above threshold (top) and of the spontaneous radiation emitted by the electron beam (bottom). After Deacon et al. (1977).

detected when the system did not satisfy the resonance condi tion; the central wavelength was 3.407 μm, and the peak output, $10^{-4}$ W. When the resonance condition was satisfied a laser oscillation took place; the spectral distribution exhibited a sharp peak as shown in Fig. 11.4 (top) with the central wavelength 3.417 μm and the peak output 7 kW. Since the transmittance of a reflecting mirror was 1.5%, it was estimated that the output power in the resonator reached 500 kW.

Such a device is called a free electron laser because it

utilizes those electrons moving in free space, though in an accelerating field, rather than those electrons in bound states of atoms or molecules. Such a free electron laser as we shall see has a special feature in that its oscillating frequency can be tuned and varied rather easily.

The experimental results described above demonstrated clearly that a relativistic electron beam injected in a wiggler field can spontaneously emit and amplify an electromagnetic wave, and is further capable of producing a laser oscillation. To analyze those experimental data quantitatively, one must clarify the nature of the radiation emitted through interaction between relativistic electrons and electromagnetic fields. We investigate those problems in the subsequent sections.

## 11.2 Electromagnetic Radiation from Relativistic Electrons

Let us begin with a review of Section 3.2. Equation (3.3) gives the intensity of an electromagnetic wave emitted from an electron when an electric field of Eq. (3.1) is applied to an electron almost stationary in the laboratory system. Carrying out a time average of $E^2$ and noting that the average energy density of an oscillating electric field is given by

$$U_E = \overline{E^2}/8\pi \quad , \tag{11.3}$$

we may write the average emission rate of radiation as

$$2\sigma_T c U_E \quad . \tag{11.4}$$

Here $\sigma_T$ is the cross section of Thomson scattering defined in Eq. (3.5).

Now turn to the case that such an electron is in motion with a relativistic velocity $\vec{v}$. We thus carry out the Lorentz transformation in the special theory of relativity. Let those physical quantities in the rest frame of the electron be expressed by symbols with primes. The electric field $\vec{E}_\perp{}'$ perpendicular to $\vec{v}$ is then

$$\vec{E}_\perp{}' = \gamma[\vec{E}_\perp + \frac{\vec{v}}{c} \times \vec{B}] \qquad (11.5)$$

where

$$\gamma = \left(1 - \frac{v^2}{c^2}\right)^{-1/2} \qquad (11.6)$$

is the **Lorentz factor** representing the ratio of the electron mass to its rest mass. To calculate the rate of emission for the electron in its rest frame, one simply uses Eq. (11.5) in Eq. (11.4), so that it is given by

$$2\gamma^2 \sigma_T c \frac{1}{8\pi} \overline{(\vec{E}_\perp + \frac{\vec{v}}{c} \times \vec{B})^2} \quad . \qquad (11.7)$$

According to the Lorentz transformation, the amount of radiation energy detected by an observer in the laboratory frame over a period $\Delta t$ corresponds to that emitted over a period $\Delta t' = \Delta t/\gamma$ in the rest frame of the electron; hence

$$2\gamma^2 \sigma_T c \frac{1}{8\pi} \overline{(\vec{E}_\perp + \frac{\vec{v}}{c} \times \vec{B})^2} \Delta t' = 2\gamma^2 \sigma_T c \frac{1}{8\pi} \overline{(\vec{E}_\perp + \frac{\vec{v}}{c} \times \vec{B})^2} \frac{\Delta t}{\gamma} \quad .$$

In its rest frame, the electron emits a dipolar radiation symmetric with respect to its position; thus the total momentum in the radiation field vanishes. Consequently, a Lorentz transformation of energy and momentum back into the laboratory frame makes the radiation energy larger by a factor of $\gamma$. This effect thus compensates exactly for the factor $\gamma^{-1}$ obtained in the transformation of period, so that the emission rate of radiation energy observed in the laboratory frame is given by

$$2\gamma^2 \sigma_T c \frac{1}{8\pi} \overline{\left(\vec{E}_\perp + \frac{\vec{v}}{c} \times \vec{B}\right)^2} \quad . \tag{11.8}$$

As Eq. (11.5) indicates, a relativistic electron perceives an enhanced acceleration effect of the electromagnetic field by a factor of $\gamma$. Consequently the intensity of radiation is enhanced $\gamma^2$ times as Eq. (11.8) shows.

Applicability of Eq. (11.8) is quite general; the cases of radiation from nonrelativistic electrons, as treated in Section 3.2, are included. The total intensities of various radiative processes can be calculated by this formula when the particular origin of the electromagnetic field $\vec{E}_\perp$ and $\vec{B}$ is specified.

For example, if the Coulomb field associated with ions or atomic nuclei is substituted for $\vec{E}_\perp$, Eq. (11.8) describes the **bremsstrahlung.** When a static magnetic field $\vec{B}$ exists in the laboratory frame, one can likewise apply Eq. (11.8) to derive the rate of **synchrotron radiation**

$$2\gamma^2 \sigma_T c \frac{B^2}{8\pi} \sin^2\psi \tag{11.9}$$

associated with a relativistic electron ($v \simeq c$). Here $\psi$ is the angle between $\vec{v}$ and $\vec{B}$. Because of this analogy, synchrotron radiation is sometimes called magnetic bremsstrahlung.

Furthermore, when $\vec{E}$ and $\vec{B}$ are the electromagnetic fields associated with an electromagnetic wave (photon), Eq. (11.8) offers a calculation for the rate of the **inverse Compton scattering** by relativistic electrons. The reason for this terminology will be explained in the next section. In the case where the electron velocity is almost parallel or anti-parallel to the direction of the incident wave, Eq. (11.8) yields

$$2\gamma^2 \sigma_T c U_R \tag{11.10}$$

where $U_R = (E_\perp^2 + B_\perp^2)/8\pi$ is the energy density of the incident wave.

## 11.3 Inverse Compton Scattering

Consider a situation in which an electromagnetic wave with frequency $\omega_1$ and wave vector $\vec{k}_1$ is scattered by a relativistic electron ($\gamma^2 >> 1$) with velocity $\vec{v}$ producing an electromagnetic wave with frequency $\omega_2$ and wave vector $\vec{k}_2$ (see Fig. 11.5). Quantum effects of scattering may be negligible as we consider only those cases where the electron energy $\gamma mc^2$ is much greater than the photon energy $\hbar\omega_1$. The incident and scattered waves satisfy the relation (3.6) with $\omega_1 = ck_1$ and $\omega_2 = ck_2$. The frequency of the scattered wave is thus calculated as

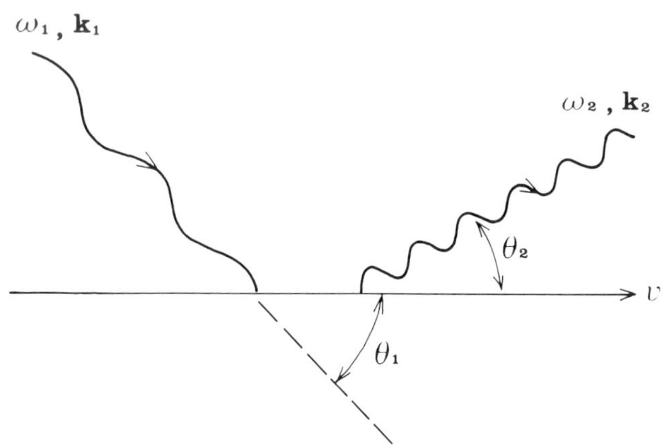


**Figure 11.5.** Inverse Compton scattering.

$$\omega_2 = \omega_1 \frac{1 - (v/c)\cos\theta_1}{1 - (v/c)\cos\theta_2} \quad . \tag{11.11}$$

Solving Eq. (11.6) for v/c and expanding the result with respect to $\gamma^{-2}$, one finds

$$\frac{v}{c} \simeq 1 - \frac{1}{2\gamma^2} \quad . \tag{11.12}$$

With the assumption $\theta_2^2 << 1$, one likewise has

$$\cos\theta_2 \simeq 1 - \frac{\theta_2^2}{2} \quad . \tag{11.13}$$

Substitution of those expansions in Eq. (11.11) yields

$$\omega_2 = 2\omega_1 \frac{1 - (v/c)\cos\theta_1}{\gamma^{-2} + \theta_2^2} \quad . \tag{11.14}$$

For those electromagnetic waves scattered into an opening angle $\theta_2$ with respect to the direction of electron motion, satisfying

$$\theta_2^2 << 1/\gamma^2 \quad , \tag{11.15}$$

Eq. (11.14) becomes

$$\omega_2 = 2\gamma^2\omega_1[1 - (v/c)\cos\theta_1] \quad . \tag{11.16}$$

The frequency $\omega_2$ is distributed between 0 and $4\gamma^2\omega_1$ depending on the value of $\theta_1$; it takes on the maximum value $4\gamma^2\omega_1$ at $\theta_1 = \pi$.

As the foregoing calculation has demonstrated, the electromagnetic wave scattered in the forward direction of the relativistic electron receives an enhancement of its frequency by a factor on the order of $\gamma^2$. Consequently, the bulk of energy in the scattered electromagnetic wave is confined in a narrow solid angle satisfying the condition (11.15); energy of the scattered photon becomes approximately $\gamma^2$ times as large as that of the incident photon. The ordinary Compton scattering, on the other hand, concerns the cases in which a relatively high-energy photon such as an X-ray collides with a low-energy electron; the photon energy decreases by an amount absorbed by recoil of the electron. The process treated in Eq. (11.16) produces an energy enhancement of the photon, in contrast to the ordinary Compton process accompanied by a decrease of the photon energy; it is thus called the **inverse Compton process.**

In a plasma with $T \simeq 10^9$ K for example, many electrons have relativistic energies. If for some reason a low-energy photon--radio wave to visible light--is supplied in such a plasma, it is probable that the photon undergoes repeated scattering with relativistic electrons and thereby receives an enhancement of its energy into the X-ray domain. Such a process is called **Comptonization** of the low-energy photon; it plays an essential part in the interpretation of physical mechanisms involved in X-ray stars.

## 11.4 Spontaneous Emission

We now return to the experiment of Section 11.1 and examine the physical process of spontaneous emission in Figs. 11.2 (a) and 11.4 (b) quantitatively on the basis of the foregoing considerations.

In the experiment a relativistic electron beam is injected along the z axis of the wiggler field Eq. (11.1) and the spectral intensity of the electromagnetic waves spontaneously excited in the beam direction is observed. Since Eq. (11.1) expresses a static magnetic field in the laboratory frame, this phenomenon may be looked upon as a kind of magnetic bremsstrahlung described by Eq. (11.9). In the rest frame of the electrons, on the other hand, the wiggler field acts effectively as a photon, so that it can also be regarded as an inverse Compton process. In any case, because $\omega_1 = 0$ and $k_1 \simeq q(\neq 0)$, Eq. (11.16) as it stands cannot be used for determination of the scattered wave; instead we begin with Eq. (3.6). Putting $\omega_1 = 0$ and $\cos\theta_1 = -1$ in Eq. (3.6), we find

$$\omega_2 = 2\gamma^2 ck_1 \tag{11.17}$$

under the condition (11.15); converting it into wavelength via $\lambda_2 = 2\pi c/\omega_2$, we then obtain

$$\lambda_2 = \pi/\gamma^2 k_1 \quad . \tag{11.18}$$

The wavelength of the generated electromagnetic wave takes on a value $1/2\gamma^2$ times as short as the wavelength $2\pi/k_1$ of the spatial variation of the magnetic field.

In the experiment of Fig. 11.1, a 24 MeV electron beam was utilized so that the corresponding Lorentz factor was $\gamma = 47$; in the experiment of Fig. 11.3, the beam energy was 43.5 MeV and thus $\gamma = 85$. To estimate the central wavelength in the spectral distribution of the emitted electromagnetic waves, we may put $k_1 = q = 2\pi/\lambda_q$ with $\lambda_q = 3.2$ cm; Eq. (11.18) then yields $\lambda_2 \simeq 7.2$ μm (Fig. 11.1) and 2.2 μm (Fig. 11.3). Those are fairly close to the experimental values--10.6 μm and 3.4 μm--respectively; there still remains a significant difference, however.

This difference can be resolved by taking an additional account of the modulation effect that the electron orbitals receive in the wiggler field. The electrons injected along the z axis start to perform spiral motion, though small in amplitude, due to the presence of the magnetic field Eq. (11.1). As a result the velocity component $v_z$ in the z direction takes on a value slightly smaller than the velocity v itself. Solving a classical equation of motion for the electron under the conditions $(\omega_H/\omega_q)^2 \ll 1$ and $\gamma^{-2} \ll 1$, one obtains

$$\frac{v_z}{c} \simeq 1 - \frac{1}{2\gamma^2}\left[1 + \left(\frac{\omega_H}{\omega_q}\right)^2\right] \qquad (11.19)$$

where $\omega_q \equiv 2\pi c/\lambda_q$ and $\omega_H \equiv e\delta B/mc$. The Lorentz factor relevant to radiation in the z direction is that associated with $v_z$. Hence, equating Eq. (11.19) with $1 - (2\gamma_*^2)^{-1}$ in analogy with Eq. (11.12), we obtain

$$\gamma_* = \gamma\left[1 + \left(\frac{\omega_H}{\omega_q}\right)^2\right]^{-1/2} , \qquad (11.20)$$

which is the Lorentz factor to be used in Eq. (11.18). With those corrections, the calculated values of the central wavelength now become $\lambda_2 \simeq 10.9$ μm (Fig. 11.1) and 3.3 μm (Fig. 11.3), showing a significantly improved agreement with the experimental values. In the subsequent treatment, the effective value Eq. (11.20) is to be used in place of the Lorentz factor.

We next calculate the total strength P of radiation emitted spontaneously by interaction between the wiggler field (11.1) and the electron beam. Total emission rate of an electron with Lorentz factor $\gamma$ is given by Eq. (11.9) where $\psi = \pi/2$ and $B^2 = \delta B^2/2$ (= mean square value of the magnetic field*). Letting

---

* The mean square value of Eq. (11.1) itself is $\delta B^2$. This strength is, however, divided equally into two components with $\cos\theta_1 = \pm 1$. Only the component with $\cos\theta_1 = -1$ is effective in the inverse Compton process under consideration; hence $B^2 = \delta B^2/2$.

$f_1(p)$ be the momentum distribution function of those electrons in z direction with normalization

$$\int dp\ f_1(p) = 1 \quad ,$$

we obtain

$$P = N\sigma_T c \frac{\delta B^2}{8\pi} \int dp\ \gamma^2 f_1(p) \tag{11.21}$$

where N is the total number of electrons pertaining to the radiation. The radiation intensities observed in the experiments were in agreement with the theoretical values.

Spectral distribution of radiation arises from various physical causes. We can, for example, think of the energy distribution and spatial spread in the electron beam and the spatial extent and homogeneity in the wiggler field. In the case of Stanford experiment, the spread in electron energies is extremely small (~ 0.05%); the spread in the observed radiation spectrum is looked upon as mainly due to the effect of size L in the wiggler field.

To carry out a spectral analysis of the field Eq. (11.1), we calculate the Fourier transformation of a function which takes on cos(qz) in the domain $-L/2 < z < L/2$, and 0 otherwise:

$$\int_{-L/2}^{L/2} dz \cos(qz)e^{-ikz} = \frac{1}{2}\int_{-L/2}^{L/2} dz[e^{i(q - k)z} + e^{-i(q + k)z}]$$

$$= \frac{\sin[(L/2)(q - k)]}{q - k} + \frac{\sin[(L/2)(q + k)]}{q + k} .$$

Since $qL/2 \gg 1$ for the field configuration used in the experiment, the ratio of the second term to the first term is on the order of $2/qL$ ($\ll 1$) over the k domain of $|q - k| \lesssim 2/L$; hence the second term is negligible. The square of this Fourier component leads to the spectral function of the field variation; it is thus proportional to

$$\Delta(k) \equiv \frac{L}{2\pi}\left\{\frac{\sin[L(q - k)/2]}{L(q - k)/2}\right\}^2 . \tag{11.22}$$

Figure 11.6 depicts the shape of Eq. (11.22). Since

$$\int_{-\infty}^{\infty} dx \frac{\sin^2 x}{x^2} = \pi$$

we may take

$$\lim_{L\to\infty} \Delta(k) = \delta(k - q) \tag{11.23}$$

on reference to the properties of the delta function (see Appendix 1).

The k in Eq. (11.22) corresponds to the $k_1$ in Eq. (11.17). The spatial fluctuation of the magnetic field with wave number

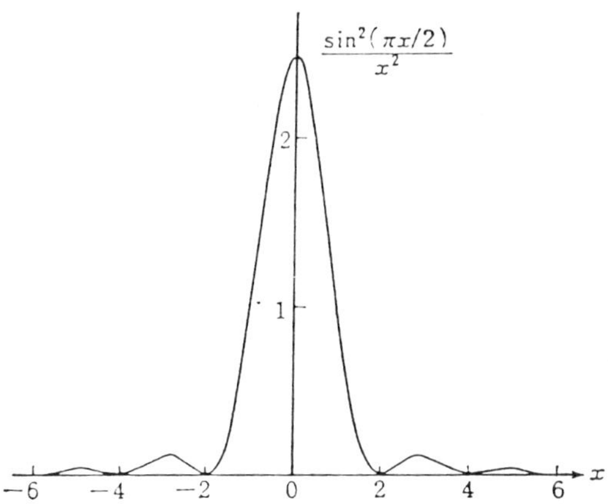


**Figure 11.6.** Shape of Eq. (11.22).

k distributes according to Eq. (11.22) and this fluctuation excites an electromagnetic wave with the frequency, Eq. (11.17), by scattering with a relativistic electron. Spectral intensity of the electromagnetic wave radiated into the opening angle of Eq. (11.15) along the z axis in the frequency interval between $\omega$ and $\omega + d\omega$ when an electron beam is injected in the wiggler field, Eq. (11.1), is thus calculated from Eq. (11.21) as*

$$\frac{dP}{d\omega} = N\sigma_T \frac{\delta B^2}{16\pi} \int dp\ f_1(p)\,\Delta\left(\frac{\omega}{2\gamma^2 c}\right) \quad . \tag{11.24}$$

When the electron beam is monoenergetic, we can replace $f_1(p)$ by a delta function $\delta(p - \gamma mc)$, so that Eq. (11.24) becomes

$$\frac{dP}{d\omega} = N\sigma_T \frac{\delta B^2}{16\pi}\,\Delta\left(\frac{\omega}{2\gamma^2 c}\right) \quad . \tag{11.25}$$

In this case, radiation spectrum of the spontaneous emission directly reflects the wave number spectral distribution of the magnetic field fluctuations. In the Stanford experiment, spectral width of the observed radiation was 0.4%; in contrast, the relative width calculated from $\Delta(\omega/2\gamma^2 c)$ is 0.3%. The principal source of the spectral width is thus seen to be the structure of the wiggler field.

---

* Use $dP/dk = P\Delta(k)$ in

$$\frac{dP}{d\omega} = \frac{1}{2\gamma^2 c} \left.\frac{dP}{dk}\right]_{k=\omega/2\gamma^2 c} \quad .$$

## 11.5 Induced Emission

When a laser light with wavelength 10.6 μm was injected in the domain of interaction between the magnetic field and the electron beam (see Fig. 11.1), the amplification characteristics of Fig. 11.2 (b) were observed. The radiative process in this case is proportional to the intensity of radiation in the interaction domain; hence it corresponds to an **induced emission.** When the rate of amplification in the interaction domain becomes substantially large, a laser oscillation can be realized through a combination with a resonator (see Fig. 11.3), as illustrated in Fig. 11.4 (a).

Calculation of the rate of induced emission for an electromagnetic wave arising from interaction between the wiggler field and an electron beam is a little more complex problem than that of the rate of spontaneous emission described in the previous section (Hopf, Meystre, Scully, and Louisell 1976; Ichimaru and Iwamoto 1978). Here, without delving into such a complicated calculation, we clarify the elementary processes involved in the induced emission and thereby infer the emission rate resulting from their combination.

Consider the rate of amplification for an electromagnetic wave with a fixed frequency $\omega_2$. The resonance condition between an electron with Lorentz factor $\gamma$ and the field fluctuation with wave number $k_1$ is given by Eq. (11.17). Those electrons with energies slightly greater than this resonance condition (larger $\gamma$) supply energy to and thereby amplify the electromagnetic wave through resonant interaction, in a process quite analogous to that described in Fig. 2.6. Those electrons with energies slightly less than the resonance condition, on the other hand, absorb energy from the electromagnetic wave and thereby decrease its amplitude. Consequently, the rate of amplification of the electromagnetic wave is proportional to the derivative of the momentum distribution function $f_1(p)$ at the resonance condition* Eq. (11.17), just as in the case of Landau damping Eq. (2.41). The rate of spontaneous emission Eq. (11.24) is, on the other hand, proportional to $f_1(p)$ itself. In order that the amplification condition of the electromagnetic wave is satisfied and that the laser oscillation is realized, there must exist more electrons in the high energy states above the resonance condition than in the low energy states below it.

Carrying out a theoretical calculation along this line, we find an expression for the rate of amplification of radiative energy per unit length of the interaction domain as (Ichimaru and Iwamoto 1978)

$$G = \frac{3}{2} n\sigma_T \frac{(\delta B)^2/8\pi}{mc^2} \mathfrak{F} \frac{\lambda_q L^2}{\gamma^3} \frac{d}{d\zeta} \left(\frac{\sin\zeta}{\zeta}\right)^2 \quad . \tag{11.26}$$

Here n refers to the number density in the electron beam,

$$\zeta = \frac{L}{2} \left(\frac{\omega_2}{2\gamma^2 c} - q\right) \quad , \tag{11.27}$$

and $\mathfrak{F}$ represents the ratio of the effective cross section of the electron beam to that of the electromagnetic wave. In the derivation of Eq. (11.26), the momentum distribution function of the electrons is assumed to be given by the delta function, as in the case of Eq. (11.25); partial integrations are carried out so that a k derivative of Eq. (11.22) is now involved in Eq. (11.26).

As may be clear from a comparison between Eqs. (11.25) and (11.26), the frequency dependence of the amplification rate G

---

* Since $p_1 = \gamma mc$ for $\gamma^2 \gg 1$, Eq. (11.17) implies $p = m(\omega_2 c/2k_1)^{1/2}$.

takes a differential form on the spectrum of the spontaneous emission. This contrast is clearly visible in the experimental data of Fig. 11.2. The maximum amplification rate over a distance L calculated from Eq. (11.26) is approximately 5.5% in the case of Fig. 11.1 and reaches 33% in the case of Fig. 11.3. The latter value may be looked upon as sufficient to induce a laser oscillation.

## Chapter 12

# Plasma Phenomena Around Neutron Stars or Black Holes

The radio astronomy commenced by Jansky's discovery of cosmic radio waves in 1931 has achieved remarkable progress due to the advancement of radio technology since then. The age of artificial satellites initiated by the successful launching of Sputnik in 1957 opened up our "eyes" of observation to the X-ray domain. The universe observed by radio waves or X-rays revealed spectacular phenomena which were beyond imagination in the conventional, visible-light observation. In 1962 a "bright" **X-ray star** was discovered in the constellation of Scorpius (Sco X-1), emitting X-rays with luminosity greater than a thousand times the total luminosity of the sun. In 1967 Hewish, Bell and their collaborators in England discovered astronomical objects regularly emitting periodic radio pulses some tens of a millisecond in width at an interval of approximately one second; the objects are called **pulsars.** In 1968 a pulsar with a pulse width of 2 ms and a period of 33 ms was discovered at the center of the Crab nebula, which is the supernova remnant of the year 1054. In 1971 an intense X-ray object with a rapid temporal variation was discovered (Cygnus

X-1); **X-ray pulsars** emitting pulsed X-rays were discovered in the constellations of Centaurus and Hercules (Cen X-3 and Her X-1). Those conspicuous astrophysical phenomena are now regarded as being produced by radiative processes in plasmas around neutron stars or black holes. A neutron star and a black hole are those states of a star expected in its final stages of evolution. An interesting problem then is how those stars which have exhausted all of their nuclear fuels can produce such conspicuous emission of radiation by plasma processes. In the closing chapter of this book, we consider those problems.

## 12.1 Pulsars

Figure 12.1 shows the record of a pulsar discovery reported in the 1968 February 24 issue of Nature (Hewish, Bell, Pilkington, Scott, and Collins 1968). This is the pulsar now known as PSR 1919 + 21;* it emits radio pulses (80.5 MHz and 81.5 MHz in Fig. 12.1) with repetition period of 1.3373 s and (average) width of 25 ms. Since then until the summer of 1980, some 328 such pulsars were discovered in the Galaxy. On average those pulsars emit pulsed radio waves with some tens of a millisecond in width and a stable period of approximately one second. We may summarize some of the typical characters of those radio pulsars (e.g., Manchester and Taylor 1977; Michel 1982) in the following way:

(1) The average period (P) and pulse shape of radio pulsars are very stable.

---

* The prefix PSR is abbreviation of "pulsar." A four-digit number following PSR indicates right ascension (in 1950.0 coordinates); a sign and two digits are added to designate degrees of declination.

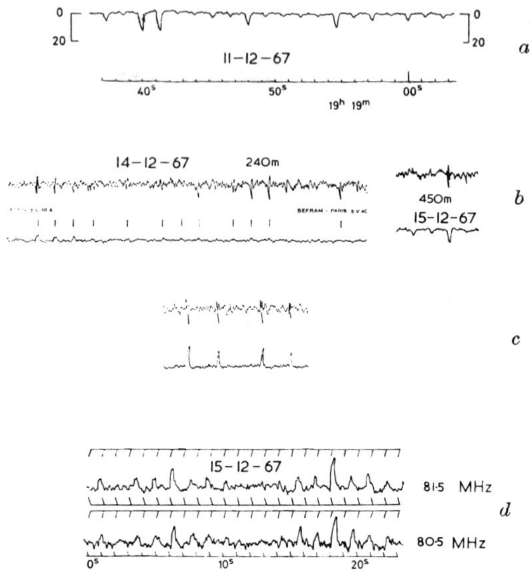


**Figure 12.1.** Record of pulsar discovery on 1967 December 11-15. After Hewish et al. (1968). Reproduced with permission. ©1968 Macmillan Journals Limited.

(2) The distribution of pulsar periods is peaked somewhere around one second; the largest period known is approximately four seconds.

(3) The repetition period of a pulsar increases gradually in time. Denoting such a rate of increase by $\dot{P}$, $\tau \equiv P/2\dot{P}$ takes on a value $10^6$ - $10^8$ years for most of the pulsars. This gives a rough estimate on the "age" of pulsars.

(4) The radio wave emitted is intense; an equivalent temperature fitted to the Rayleigh-Jeans law reaches $10^{21}$ - $10^{30}$ K.

(5) Average pulse width is approximately proportional to the period; the pulse width ranges 4° - 20° when the pulse period is taken to be 360°.

(6) With a few exceptions, the pulsar radio waves are weakly polarized (Manchester, Taylor, and Huguenin 1975). When a polarization is observed, a linear polarization is predominant and in some cases circular polarizations are observed.

(7) Looking at individual pulses with a resolution of ~ 1 ms, one can distinguish various subpulses with widths 2° - 5° (Taylor, Manchester, and Huguenin 1975). For some pulsars those subpulses exhibit a systematic drift motion across the pulse window.

(8) Those observed pulsars belong to the Galaxy; their distances from the earth are on the order of a thousand light years.

The characterizations described above concern common features known to many of the observed pulsars. As for the notorious pulsar found in the Crab nebula (PSR 0531 + 21), we add the following remarks: Its period P = 33.098 ms is one of the shortest among the known pulsars; $\dot{P} = 4.227 \times 10^{-13}$ is the largest. The calculated value of $\tau$ is 1200 years, which roughly corresponds to the known fact of its birth in the year 1054. The Crab pulsar is a young and active pulsar, emitting not only strong radio pulses but also pulsed visible light and X-rays.

## 12.2 Rotating Magnetic Neutron Stars

When the discovery of pulsars created a wake of a great sensation, a first question naturally asked was: What is the pulsar? Candidates for the astronomical objects which can sustain variations with such a short characteristic time ranging a few milliseconds to a few seconds are confined to those extremely compressed astronomical objects such as a white dwarf and a neutron star, expected in the final stages of stellar evolution. As for the characteristic motion, the stellar rotation and vibration and the orbital motion of a binary system were considered; the last one among them was soon discarded because of a lifetime consideration due to emission of the gravitational waves.

The **white dwarf** (e.g., Shapiro and Teukolsky 1983) is an astronomical object in which the gravitational self-contraction is balanced by the pressure arising from the Fermi energy of the degenerate electrons. Typically its mass is taken as approximately the solar mass ($M_\odot \simeq 2 \times 10^{33}$ g) and the radius approximately one percent of the solar radius ($R_\odot \simeq 7 \times 10^5$ km). This corresponds to compressing a solar mass into the volume of the earth, so that the mass density in its central domain reaches $10^6$ - $10^8$ g/cm$^3$. A calculation on the basis of such a white dwarf model reveals that the characteristic period of the radial oscillation in its fundamental mode cannot become smaller than 2 s; a consideration on the basis of the escape velocity of matter from the stellar surface indicates that a theoretical lower limit of the rotational period is approximately 1 s. Those numbers were somewhat inconvenient for explanation on the bulk of the observed pulsars. It was also pointed out that none of the observed pulsars were connected to the known white dwarfs. For the reasons stated above the white dwarfs have almost been rejected as candidates of the pulsars.

The **neutron star** (e.g., Shapiro and Teukolsky 1983) is a final state of a star which may be realized in the supernova remnant; it is even more compressed than the white dwarf. Typically one assumes a neutron star of the solar mass with a radius of approximately 10 km; the density in its central domain reaches $10^{14}$ - $10^{15}$ g/cm$^3$. Period for the radial vibration of such a neutron star gives $10^{-3}$ - $10^{-4}$ s, which appears too small to account for the pulsar observation. The lower limit of the rotational period, however, becomes $10^{-2}$ s, which is inadequate to explain pulse repetition periods of pulsars including the Crab pulsar. Observed connection between pulsars and supernova remnants such as in the case of Crab nebula have been regarded as a positive evidence for the neutron star model of pulsars.

Although by the method of elimination as stated above, many investigators now appear to agree that the pulsars originate from rotating neutron stars. That the signals from PSR 0950 + 08 contain rapid impulses of 175 μs and are frequently accompanied by variations of ~ 10 μs (Rickett, Hankins, and Cordes 1975) is also a favorite piece of evidence for the neutron star hypothesis. The total extent of the radio source would at most be the distance over which the radio wave propagates during the time of a rapid variation observed. Distances over which the radio wave propagates during 10 - 175 μs are 3 - 50 km. These are about the same order of magnitude as the typical neutron star radius of 10 km, but are substantially smaller than the white dwarf radius.

It is expected that a neutron star is accompanied not only by a rotational motion but also by a strong magnetic field. Consider a situation in which a star similar to the sun with an average surface magnetic field of $10^2$ gauss and a rotational velocity of $10^{-7}$ rad/s is compressed by a factor of $10^{-5}$ in radius and thereby becomes a neutron star. Assuming also that both the total magnetic flux and the total angular momentum are conserved in the process of contraction, we estimate that the resulting neutron star would have a surface magnetic field of approximately $10^{12}$ gauss and a rotational angular velocity of $10^3$ rad/s.

This value of rotational velocity, of course, far exceeds the repetition frequency of pulsars now observed. The rotational energy, however, is gradually lost in time and the neutron star rotates at a reduced velocity accounting for the present observation. This is manifested in the observed increase in the pulsar period ($\dot{P} > 0$).

The moment of inertia I of a neutron star may be estimated from the typical parameters mentioned earlier. Assuming the mass $M \simeq 10^{33}$ g and the radius $R \simeq 10^6$ cm, one finds $I \simeq MR^2 \simeq 10^{45}$ g $\cdot$ cm$^2$. The rotational energy

$$E_R = \frac{1}{2} I\Omega^2$$

associated with the spinning motion of the neutron star with the angular velocity $\Omega = 2\pi/P$ decreases at a rate, $\dot{E}_R = I\Omega\dot{\Omega}$. For the Crab pulsar the magnitude of this $\dot{E}_R$ is computed to be $\sim 10^{39}$ erg/s, which turns out to be sufficient to supply all the luminosity of the Crab nebula, $\sim 4 \times 10^{38}$ erg/s.

This observation provided a solution to the outstanding puzzle concerning the activity of the Crab nebula. The Crab nebula is the remnant of the supernova explosion in the year of 1054, and now spreads over a spatial extent of several light years. From the region, a wide spectrum of radiations ranging from the radio waves to the $\gamma$-rays are emitted. The observed radiation characteristics clearly indicate that those emissions are created not thermally but by a nonthermal mechanism such as the interaction of relativistic electrons with magnetic fields as treated in the preceding chapter. Calculated lifetimes of those high-energy electrons are far shorter than the age of the Crab nebula, so that those electrons must be supplied continuously by some mechanism. Since the existence of a pulsar in the central part of the Crab nebula is confirmed, it would be natural to assume that the energy released by the pulsar provides an acceleration mechanism for electrons. It might give us a great and elevated feeling to imagine that a heavy "toy top" with a radius of 10 km spinning at a high speed of 30 revolutions per second releases its rotational energy gradually and thereby "energizes" the entire nebula surrounding it over an extent of several light years. It is now a widely accepted point of view that the rotational energies of the neutron stars provide the energy source of activities not only for the Crab pulsar but also for all other pulsars.

It is natural then to suppose that the strong magnetic field associated with a neutron star plays an essential part in the process of converting rotational energy to radiation energy. For example, when the axis of the magnetic field is tilted by a certain angle with respect to the axis of rotation (see Fig. 12.2), the rotational energy is consumed by magnetic dipole radiation. Assuming this mechanism, one can estimate the magnitude of the surface magnetic field from the measured values of P and $\dot{P}$. It is interesting that strengths of the magnetic field estimated in this way remain around $10^{12}$ gauss for many pulsars. The frequency of such a magnetic dipole radiation, however, is $P^{-1}$; this mechanism by itself would not produce pulses. Hence the magnetic dipole radiation is not directly connected to the observed pulsar phenomenon.

Let us proceed with consideration of pulsar radiation mechanisms by assuming a fundamental model of a rotating neutron star with a surface magnetic field of approximately $10^{12}$ gauss (e.g., Manchester and Taylor 1977; Michel 1982). Combining the observed strength of the radio waves and an estimate of pulsar distance, we can infer the energy density of radiation field in the vicinity of a pulsar. This energy density must be substantially smaller than the energy density of the plasma itself which produces the radiation field. Let us now

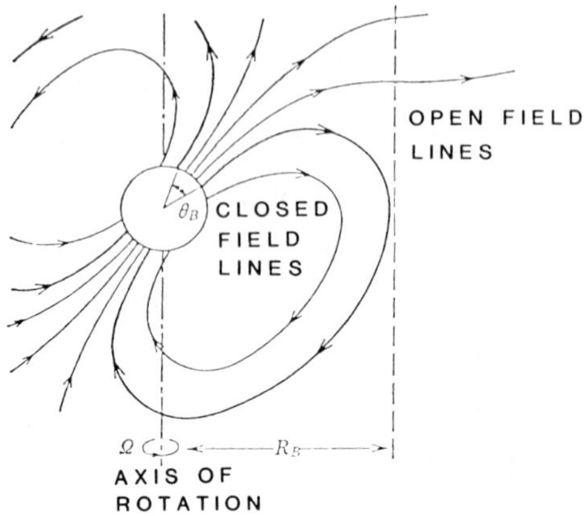


**Figure 12.2.** Rotating magnetic neutron star.

recall the stability of the average pulse shape mentioned as item (1) in Section 12.1. To be able to emit stable signals, the emitter--a plasma in this case--must have a "firm" construction with a mechanical stability. The only conceivable source to provide such a stability is the strong magnetic field associated with a neutron star. For a magnetic field to be able to firmly control the plasma behavior, the energy density of the magnetic field must also be considerably larger than that of the plasma. Examination of those conditions numerically leads one to accept a strong possibility that the radio waves are emitted in the domain close to the surface of a neutron star.

A simple mechanism to recreate the observed radio pulses by a rotating star is to assume a searchlight beam of radio wave around the axis of rotation. In this model, the opening angle of the searchlight beam is directly related to the effective pulse width. It may then be natural to assume that the magnetic-field configuration makes the determining factor for the opening angle of the beam as well.

Some of the magnetic lines of force streaming out from the surface of a rotating star may return to the stellar surface, but those starting from the vicinity of the magnetic poles would stream out into interstellar space (Fig. 12.2). The former constitute closed magnetic lines of force, while the latter form open magnetic lines of force. Those plasmas confined in the space occupied by closed magnetic lines of force cannot be effectively accelerated in the force field produced by the rotating magnetic field. The accelerating mechanisms are effective to plasmas in the open magnetic-field domain, create high-energy charged particles, and thereby lead to emission of the conspicuous radiations. Consequently, the boundary between the open magnetic field domain will determine the opening angle of the searchlight.

Let $R_B$ be the largest radius where the closed magnetic lines of force can reach. Assuming a dipolar magnetic field, we may calculate the opening angle of the searchlight as

$$\theta_B = 3(R/R_B)^{1/2} \tag{12.1}$$

where R is the radius of the neutron star. Two theoretical possibilities have been suggested to determine $R_B$. One is to take

$$R_L = c/\Omega \tag{12.2}$$

where the rotational velocity becomes equal to the velocity of light. The other is to take $R_c$ where the gravitational force balances with the centrifugal force (Roberts and Sturrock 1973), that is,

$$\frac{GM}{R_c^{\,2}} = \Omega^2 R_c \quad . \tag{12.3}$$

Here $G = 6.6720 \times 10^{-8}$ erg • cm • $g^{-2}$ is Newton's gravitational constant and M is the neutron star mass. This radius being an order of magnitude smaller than $R_L$ gives a slightly larger estimate of $\theta_B$. According to the model of Fig. 12.2, the $\theta_B$ value determines the relation between the pulse width and its period. Numerical comparison with observational data has shown that an estimate of the searchlight opening angle in either model can be made fairly consistent with the observation.

Following the idea explained above, one can define the radiative domain of a pulsar and its outer boundary. The remaining problems for a pulsar model are to apply an accelerating field onto the plasma in this radiative domain, to predict the plasma behavior and the character of the emitted radiation, and finally, to compare the prediction with the observational data. These are in fact difficult problems. The reason for the difficulty stems in part from the fact that we cannot definitely specify how those plasmas are produced, not to mention the plasma properties. Without a solution to such a problem of plasma production, it would be out of the question to analyze the radio emission mechanisms in the plasma.

Sturrock (1971) and later Ruderman and Sutherland (1975) proposed a theoretical model accounting for radiation mechanism of pulsars with inclusion of the plasma production problem. To a degree it is analogous to the processes found in ordinary gaseous discharges. As for the method of plasma production, one evokes the electron-positron pair production arising from interaction between the $\gamma$-rays and magnetic field, in place of the electric discharge of atoms and molecules in gases. The threshold energy $2mc^2$ for pair production thus plays the part of the ionization potential. Let us outline the explanation proposed for pulsar phenomena on the basis of this model.

A neutron star has a finite electric conductivity because it contains a substantial amount of conduction electrons and protons. The calculated values of the electric conductivity for the internal state of a neutron star, in fact, take on considerably large values, so that a neutron star may be regarded almost as a perfect conductor. Generally, in the rotating magnetic neutron star, the stellar matter would move relative to the magnetic field; this motion in turn would exert a force onto the stellar matter. In a perfect conductor a net force acting on a charged particle should vanish; a space-charge distribution is thus created inside the star just to cancel the magnetic force. In other words, a rotating magnetic star polarizes itself electrically and thereby produces an internal electric field.

Electric polarization of a star also induces a potential difference in the space exterior to the star. The maximum amount of the potential difference created between the star and the interstellar space along the magnetic lines of force reaches $\Omega\Phi/2\pi c$, where $\Phi(\simeq \pi\theta_B{}^2R^2B/4)$ is the total number of the open magnetic lines of force in Fig. 12.2. The bulk of such a potential difference is moreover concentrated near the stellar surface up to an altitude on the order of the stellar radius. This potential difference would naturally increase with the rotational velocity $\Omega$ of the star; for an ordinary pulsar with $P \simeq 1$ s it takes on a value of approximately $10^{12}$ V.

Suppose that an electron (or a positron) is placed above the neutron star surface near a magnetic pole. Since the electric field is extremely strong, such an electron would be accelerated immediately to a super-relativistic energy ($\gamma \gg 1$ for a particle with energy $\gamma mc^2$). The relativistic electron traveling along a magnetic line of force with a finite radius of curvature emits a photon in its direction of motion with a characteristic energy determined by the radius of curvature and the $\gamma$ value. This radiation process is essentially the same as those considered in Section 11.3. When the energy of such a photon exceeds $2mc^2$, a possibility arises of creating an electron-positron pair through interaction of the photon with adjacent magnetic lines of force. The mean free path of a photon for the pair production is a function of the strength of magnetic field, the curvature, and the photon energy; it tends to decrease as each of those quantities increases. When the effective length of the accelerating electric field is comparable to such a mean free path, a pair production takes place; the electron and positron so produced are then accelerated separately by the electric field and thereby emit photons; those photons then produce electron-positron pairs... . An avalanche thus takes place and a "spark" discharge may be maintained. Consequently we may estimate the effective length of the electric field as comparable to the mean free path of a photon for the pair production; its magnitude increases as the rotation velocity and the magnetic field decrease.

If a spark discharge of this nature is maintained in the space near the surface of a magnetic pole, an electron-positron plasma can thus be produced. Since an electron and positron carry electric charges of opposite sign, they produce a counter-streaming flow in the electric field. When such a counter stream of charged particles exists in a plasma, the plasma oscillation becomes unstable (Buneman 1959) and a spatial bunching of charged particles may take place; such a phenomenon is called a **two-stream instability** of a plasma. We considered an instability of plasma oscillation in connection with Eq. (2.41) in Section 2.7. The two-stream instability is a quite analogous phenomenon, although it cannot be directly derived from Eq. (2.41).

The bunches of charged particles travel along magnetic lines of force and thereby emit electromagnetic waves by the curvature acceleration. If the sizes of bunches are smaller than those wavelengths in the radio domain, the N charged particles forming a bunch may emit electromagnetic waves with coherent phases. The strength of electromagnetic radiation is proportional to the square of the electric charge. When the N charged particles oscillate coherently, they emit electromagnetic waves with a strength N times as much as in the case of incoherent radiation. Let us recall that we have already treated an analogous effect in Section 7.4 in connection with the stopping power of a plasma against ion clusters. Since N can take on an extremely large number, the high effective temperatures observed for pulsars [item (4) in Section 12.1] may be accounted for in terms of such a bunching mechanism.

As the rotation is slowed down and the effective length of the accelerating electric field increases, the efficiency of accelerating electrons and positrons decreases; it may eventually turn out that a high-energy photon necessary for the pair production cannot be created. In these circumstances, the spark discharge, and thus the electron-positron plasma cannot be maintained; even if such a neutron star has a substantial strength of magnetic field, it cannot be observed as a pulsar. This is said to be the reason for not finding pulsars with a long period.

The foregoing is a scenario explaining the mechanism of pulsar phenomena. Various other possibilities and alternatives have been considered and examined (e.g., Manchester and Taylor 1977; Michel 1982). It appears that much more work is necessary before a final answer is reached. As the foregoing example has illustrated, we may certainly say that plasma physics should play an essential part in clarification of the pulsar phenomena.

## 12.3 X-Ray Pulsars

Except for a few examples (e.g., PSR 0531 + 21), all the pulsars treated in the previous section are radio pulsars. In 1971, X-ray pulsars were discovered in the constellations of Centaurus and Hercules, and named Cen X-3 and Her X-1 (Giacconi et al. 1971; Tananbaum et al. 1972b). Figure 12.3 shows the X-ray signal from Her X-1. The pulse repetition period is 4.84 s for Cen X-3, and 1.24 s for Her X-1.

It was also confirmed that those X-ray pulsars are members of binary stars. The binaries perform Keplerian orbital motion

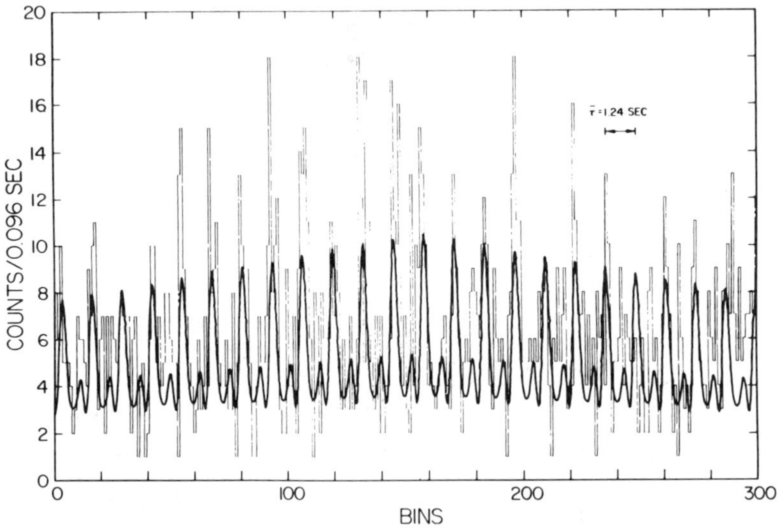


**Figure 12.3.** The counts accumulated in 0.096-second bins from Hercules X-1 during the central 30 seconds of a 100-second pass on 1971 November 6. The heavier curve is a minimum $\chi^2$ fit to the pulsations of a sine function, its first and second harmonics plus a constant, modulated by the triangular response of the collimator. After Tananbaum et al. (1972b).

with respect to their centers of gravity. The period of such an orbital motion was measured to be 2.1 days for Cen X-3 and 1.7 days for Her X-1. Note that those are far shorter than the orbital period of the earth (approximately 365 days), for example.

The orbital motion was confirmed by the fact that the pulse arrival times were not evenly distributed (Fig. 12.4). This is the Doppler effect treated in Section 3.3. While an X-ray pulsar in the binary is approaching toward us we observe interpulse separations shorter than the average; while it is receding from us the pulses would be observed with wider separations. The orbital data for the X-ray pulsar can thus be deciphered from observation of the pulse arrival times. The deviations of the pulse arrival times from the average, shown in Fig. 12.4, follow a sinusoidal curve; it can be said that the X-ray pulsar performs an almost circular orbit with negligible eccentricity.

Another remarkable feature in Fig. 12.4 is that the X-ray

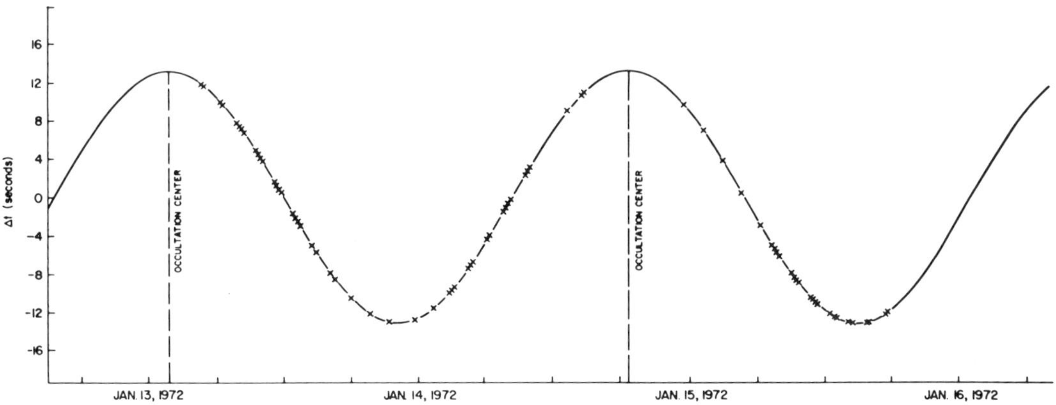


**Figure 12.4.** The difference Δt between the time of occurrence of a pulse and the time predicted for a constant period is plotted as a function of time. After Tananbaum et al. (1972b).

pulses are unobservable at certain phases of the orbital motion. This implies an "eclipse" when a small X-ray star travels behind a giant star with a radius several times that of the sun. This eclipse provides another piece of observational evidence that the X-ray pulsar is a member of a binary system.

Analysis of the data concerning the pulse arrival times reveals the following orbital information on the X-ray star. Because the orbital period is short, the distance between the binary components takes on a correspondingly small value of approximately $10^7$ km in light of Kepler's third law. Such a system is called a **close binary system**; the distance cited above is comparable even with the solar radius.

Let us now take a system of coordinates fixed onto such a binary system, and consider a force acting on a test mass placed in such a system. It is first of all attracted by the gravitational force toward the center of gravity of each star. It is also affected by the centrifugal force arising from the orbital motion of the binary system around its center of mass; it is thus repelled from this center of mass. When the test mass is located far away from the binary system, the centrifugal force is the dominant effect; the test mass will be ejected. When it is placed close to the binary system, on the other hand, the gravitational attraction wins.

It is possible to define an effective potential and the associated equipotential surfaces for the combined field of the centrifugal force and the gravitational force. As Fig. 12.5 shows, the **Roche limit** is defined as the smallest of those equipotential surfaces which enclose both stars. The effective potential takes on a maximal value at L along the line ab and a minimal value at L along the line cd; the point L thus makes a saddle point for the effective potential. If the right-hand star in Fig. 12.5 has a size comparable to or greater than the

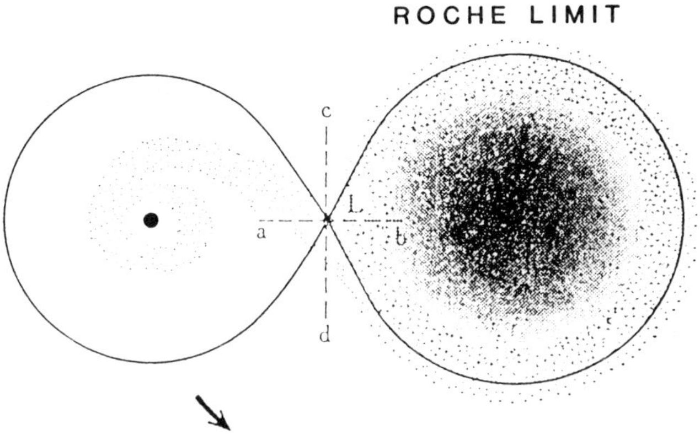


**Figure 12.5.** Roche limit and accretion in the close binary system.

Roche limit then the atmospheric matter of the star flows over the Roche limit through the saddle point L and falls into the gravitational domain of the left-hand star. When the left-hand star is a dense degenerate star such as a neutron star (or a black hole), a sizable amount of gravitational energy is liberated as the over-flowing matter falls onto the stellar surface; a copious emission of X-ray is thus expected.

The process described above is called **accretion.** Let us calculate the amount of energy liberated through an accretion process. We denote the radius of the accreting star by R, the mass by M, and the rate of mass accretion per unit time by $\dot{M}$. Since the accreting matter eventually sticks to the star, it corresponds to the rate of increase in the mass of the left-hand star in Fig. 12.5. The rate of gravitational energy liberated is thus calculated as

$$\frac{GM\dot{M}}{R} = 8.4 \times 10^{36}\left[\frac{M}{M_\odot}\right]\left[\frac{\dot{M}}{10^{-9}M_\odot/\mathrm{yr}}\right]\left[\frac{R}{10\mathrm{km}}\right]^{-1} \quad (\mathrm{erg/s}) \quad . \qquad (12.4)$$

Although we cannot predetermine the magnitude of $\dot{M}$, we also know that the sun ejects substantial amount of its mass in the form of solar wind. Since such an ejection of stellar matter into its surrounding space is rather a common happening, it may be natural to assume a mass transfer at a rate of $\dot{M} = (10^{-9} - 10^{-8}) M_{\odot}/\mathrm{yr}$. Taking $M = M_{\odot}$ and $R = 10$ km corresponding to a neutron star, we compute from Eq. (12.4) the rate of releasing the gravitational energy at $10^{37} - 10^{38}$ erg/s. Since the total luminosity of the sun is approximately $4 \times 10^{33}$ erg/s, the foregoing computation points to a possibility of devising an X-ray star "brighter than a thousand suns" through accretion process onto a neutron star in a close binary system.

Such an accretion of matter in a close binary system is the most popular among those fundamental models accounting for X-ray pulsars such as Cen X-3 and Her X-1 (Pringle and Rees 1972; Davidson and Ostriker 1973; Lamb, Pethick, and Pines 1973; Shapiro and Teukolsky 1983). As it falls in the gravitational field of the neutron star, the matter may form an accretion disk, to be explained in the next section. As in the case of the radio pulsar, we may assume that the accreting neutron star is strongly magnetized. The flow direction of the accreting plasma is controlled by the stellar magnetic field in the vicinity of the neutron star; the plasma thereby falls in the vicinity of the magnetic poles, forming an accretion column. Through the accretion process the plasma is strongly heated and thereby emits X-rays. The strength of X-rays observed from the earth is modulated by the stellar rotation; periodic pulse signals as shown in Fig. 12.3 are produced.

From the point of view of plasma physics, the setting around an X-ray pulsar differs significantly from that around a radio pulsar. In the case of the radio pulsar, we had to solve the problem of plasma production at the onset. In the accretion model of an X-ray pulsar, on the other hand, the origin and fundamental properties of the plasma are more or less given. The mass accretion rate can be determined from the X-ray luminosity according to Eq. (12.4); hence the plasma density can be estimated. The plasma temperature can also be inferred from the energy spectrum of the observed X-rays. We thus have a fairly dependable estimate on two of the fundamental quantities--density and temperature--describing the plasma properties.

On the basis of those considerations, one may proceed to analyze hydrodynamic behavior of the accreting plasma, its interaction with the stellar magnetic field, and heating and X-ray emission in the vicinity of the magnetic poles. The problems that plasma physics has to face are then to elucidate the mechanisms of the X-ray pulsars through comparison of the observational data with the theoretical predictions.

## 12.4 Black Hole Model of Cygnus X-1

In 1971 another X-ray star, Cygnus X-1, attracted attention of many investigators (Oda et al. 1971; Rappaport, Doxsey, and Zauman 1971; Schreier et al. 1971). The X-ray signal from Cyg X-1 exhibited violent short-term variabilities as shown in Fig. 12.6, although it did not contain a well defined periodic component as in the cases of Cen X-3 and Her X-1. The X-ray intensity varied with a time constant less than a second; with a detailed measurement X-ray bursts of a few hundred microseconds were observed.

It was confirmed also that Cyg X-1 is a member of a close binary system with the orbital period of 5.6 days. As the companion of Cyg X-1, a star HDE 226868 was optically identified; its absorption lines showed Doppler effect due to the orbital motion. It was also observed that the emission lines

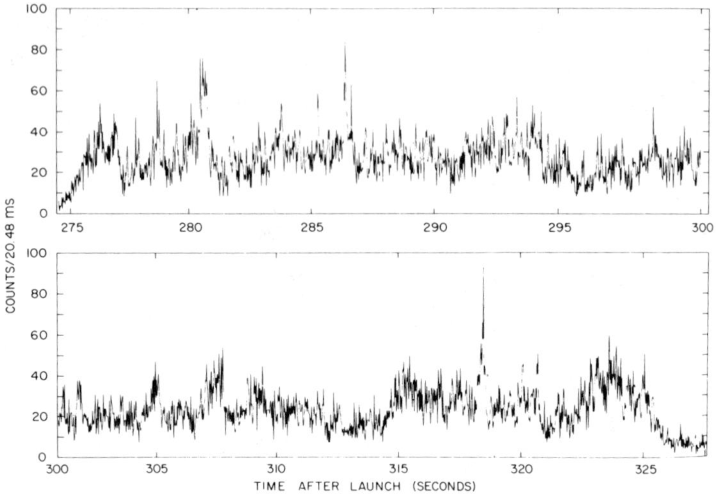


**Figure 12.6.** The entire exposure to Cygnus X-1 as a function of time after launch on October 4, 1973. The count rates are binned every 20.48 ms. After Rothschild et al. (1974).

from the matter accreting onto Cyg X-1 exhibited the Doppler effect which varied in opposite phases.

Figure 12.7 shows salient features observed in the energy spectrum of X-rays (Tananbaum et al. 1972a). The abscissa and ordinate represent the X-ray energy and the count numbers of X-ray photons (per unit area per unit time), both plotted in logarithmic scales. Until early April of 1971, the Cyg X-1 spectrum was that shown by the kink line in Fig. 12.7. Since the energy emission rate is given by the product between energy per photon and the count number, this kink spectrum implies a two peak structure--in the vicinity of a few keV and a few tens of a keV--for the energy emission rate. Around early April of 1971, however, the spectrum changed into a shape of the straight line in Fig. 12.7 during a time period of approximately one day; the peak in the vicinity of a few keV disap-

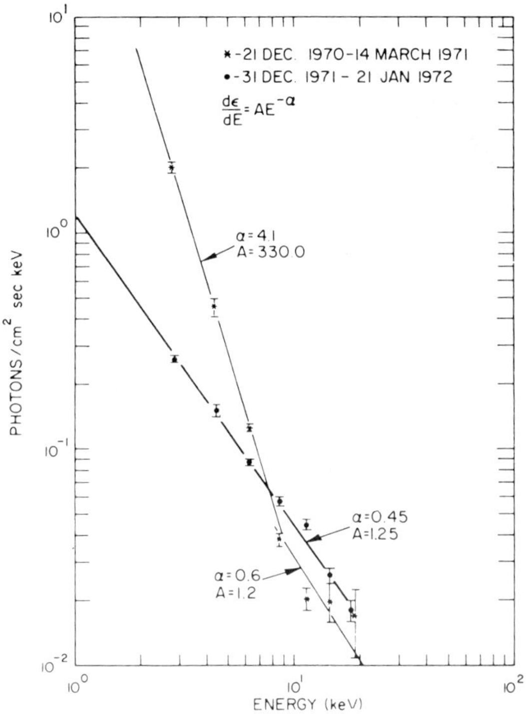


**Figure 12.7.** The average spectra for Cygnus X-1 from before and after the 1971 March and April transition. The spectra are plotted in photons $(cm^2 \cdot s \cdot keV)^{-1}$ versus energy, and various power law fits are labeled with the energy index and normalization coefficient. After Tananbaum et al. (1972a).

peared (Tananbaum et al. 1972a). This mode of the spectrum remained until April 1975, when the spectrum returned gradually to the two peak structure over several days in late April (Holt, Boldt, Kaluzienski, and Serlemitsos 1975). In early May Cyg X-1 again assumed a single peak mode with a transition time of approximately one day (Sanford et al. 1975). After that, until today, several transitions between those two modes have been observed.

To summarize the observed features of Cyg X-1, we may note an intense X-ray emission ($\gtrsim 10^{37}$ erg/s) with rapid time variabilities (Fig. 12.6), existence of two "modes" in the energy spectrum (Fig. 12.7), and transitions between those two modes.

It has been postulated that Cyg X-1 is a black hole in a close binary system. For the energy source of the X-ray emission, an accretion process analogous to that described in Fig. 12.5 has been assumed. A black hole is a state of a star compressed further beyond the state of a neutron star (e.g., Weinberg 1972; Shapiro and Teukolsky 1983). The associated gravitational field is so strong that even a photon is trapped and absorbed within a certain critical radius around a black hole.

One of the principal reasons for postulating a black hole derives from the mass estimate for Cyg X-1. We have noted that Cyg X-1 and HDE 226868 constitute a close binary system. The character of the orbital motion can be deduced from combinations of the observed Doppler shifts in the absorption lines and Kepler's laws. From the spectral type of HDE 226868 one derives a relation between its radius and mass. Since an accreting plasma flow has been observed, the companion star may be assumed to fill almost its own Roche limit. Combinations of those conditions enable one to determine a possible range for the mass M of Cyg X-1 as $9\ M_\odot \lesssim M \lesssim 18\ M_\odot$ (Bolton 1975).

Since Cyg X-1 emits X-rays with a rapid short-term variability (see Fig. 12.6), it is natural to assume that its main body is a compact star. A white dwarf or a neutron star has an upper limit in its mass for a stable existence. For the white dwarf it is the **Chandrasekhar limit** and takes on a value approximately 1.4 $M_\odot$. The theoretical calculation of the upper limit for the mass of a neutron star still involves certain elements of uncertainty, but it appears that the upper limit is likely to fall around 3 $M_\odot$. At any rate, for a stellar mass greater than 9 $M_\odot$ neither a white dwarf nor a neutron star can exist stably; the only conceivable object remaining is a black hole.

Let us then describe a scenario as to how the accreting plasmas behave and emit X-rays in a close binary system containing a black hole (Thorne and Price 1975; Eardley, Lightman, and Shapiro 1975; Shapiro, Lightman, and Eardley 1976; Ichimaru 1977). The numbers cited are the model parameters adopted (Ichimaru 1977) for Cyg X-1.

Let r be the distance from the accreting black hole. General relativistic effects become important for $r \lesssim 10^7$ cm. The Roche limit is located around $r \simeq 10^{12}$ cm; stellar matter (plasma) overflows through the point L in Fig. 12.5 and falls into the gravitational domain of the black hole. Because of the orbital motion in the binary system, the accreting plasma carries angular momentum with respect to the black hole. Those plasmas flowing with angular momentum in the gravitational field tend to form a disk in the plane perpendicular to the rotational axis (see Fig. 12.8). This is an **accretion disk** (Pringle and Rees 1972; Shakura and Sunyaev 1973; Novikov and Thorne 1973).

The thickness 2h of an accretion disk can be calculated from the condition for equilibrium between the expanding force ($\sim P/h$) due to the pressure P and the contracting force ($\sim GM \rho_m h/r^3$) due to the gravitational attraction of the black hole, i.e.,

$$2h \simeq 2r\left(\frac{P}{\rho_m}\right)^{1/2}\left(\frac{r}{GM}\right)^{1/2} \quad . \qquad (12.5)$$

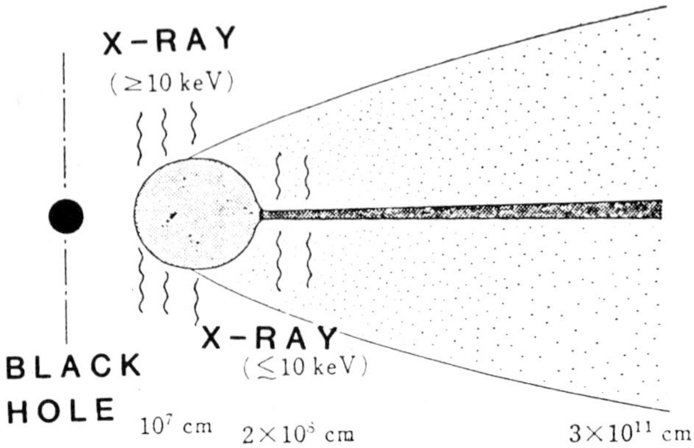


**Figure 12.8.** Schematic cross-sectional view of accretion disks around a black hole, showing a possible bimodal behavior.

Here $\rho_m$ refers to the mass density of the plasma and

$$v_K = (GM/r)^{1/2} \tag{12.6}$$

is the Keplerian rotational velocity of the disk. Such an accretion disk may be formed at distances, $r \lesssim 3 \times 10^{11}$ cm.

The angular momentum per unit mass $rv_K$ is proportional to $r^{1/2}$. For the plasma to be able to fall onto the black hole, excess amount of the angular momentum should be released outward. Viscosity of the plasma participates in the process of such an angular momentum transfer (Eardley and Lightman 1975; Ichimaru 1976). Viscosity also implies frictional effects and thus acts to heat the plasma.

Numerical analysis of the angular momentum transfer, the viscous heating and the resulting motion of the plasma indicates a possibility of two different modes in the structure of the accretion disk, as illustrated schematically in Fig. 12.8 (Ichimaru 1977). Distinction between those two modes depends on the boundary conditions at $r \simeq 3 \times 10^{11}$ cm where the disk is formed.

When relatively high-density plasma is supplied at the boundary ($\dot{M} \gtrsim 3 \times 10^{-8}\ M_\odot/yr$), a low-temperature disk starts to form owing to high efficiency in the thermal emission of radiation; heat generated by viscosity is efficiently carried away by radiation. As $P/\rho_m$ is proportional to the temperature, h/r determined from Eq. (12.5) takes on a small value; hence a thin disk is formed.

As the plasma disk gradually falls toward the black hole, the gravitational energy is released and the temperature rises. The low energy peak (~ a few keV) in the kinked spectrum of Fig. 12.7 comes from thermal radiation in this vicinity ($r \gtrsim 2 \times 10^8$ cm).

As the plasma falls further toward the black hole, such a thin disk becomes thermally unstable in the vicinity of $r \simeq 2 \times 10^8$ cm. For, if the disk is heated and expanded, the efficiency of thermal radiation decreases; the thermal energy due to viscous heating cannot be efficiently dissipated and the plasma temperature goes up futher and an instability thus develops. As a consequence, the disk plasma is abruptly transformed into a high-temperature expanded state; its electron temperature reaches $\sim 10^9$ K. In this domain the inverse Compton processes discussed in the previous chapter become effective, and high-energy X-rays with a peak around a few tens of a keV are emitted. The radiative processes in the mode represented by the kink line in Fig. 12.7 may be accounted for in terms of the evolution of accretion disk as described above.

When a relatively low density plasma ($\dot{M} \lesssim 3 \times 10^{-8}\ M_\odot/yr$) is supplied in the vicinity of $r \simeq 3 \times 10^{11}$ cm, on the contrary, the accretion disk follows an evolutional process qualitatively different from that described above. The rate of bremsstrahlung depends on the frequency of interparticle collisions, which in turn is proportional to the square of density. A low density plasma with scarce interparticle collisions has a low efficiency of radiation; hence, the viscous dissipation effect goes directly into heating of the plasma. The accreting plasma begins with a high-temperature expanded state. Although the plasma temperature is high, the X-ray emission is weak in the beginning because of the low mass density.

Such an accretion disk, however, joins into the high-temperature expanded plasma state of the previous evolutional scenario in the vicinity ($10^7$ cm $\lesssim r \lesssim 2 \times 10^8$ cm) of the black hole; it thus produces an X-ray spectrum with a peak around a few tens of a keV. We may thus interpret the mode represented by the straight line in Fig. 12.7 as originating from such an accretion process.

The high temperature plasma in the vicinity of the black hole ($r \gtrsim 10^7$ cm) moves so fast that the plasma may well be in a magnetohydrodynamically turbulent state. Consequently, the observed millisecond bursts may be accounted for in terms of violent motions associated with turbulence; the rapid X-ray variability characteristic of Cyg X-1 may thus reflect the special features in the accreting plasma onto the black hole.

A black hole model has given a consistent explanation for the bimodal characteristics of the Cyg X-1 spectrum. Combining these facts with the mass estimation of Cyg X-1, we may assume it highly probable that Cyg X-1 is in fact a black hole. Detailed quantitative analyses were carried out on the black hole model of Cyg X-1 with inclusion of the nature of accreting plasmas; we have outlined some of the principal features above.

The black hole was predicted originally as a theoretical singular point in the field of general relativity. It now appears highly probable through analyses of the observational data on Cyg X-1 that the black hole really exists in the universe. Plasma physics has played an important part in providing a connecting bridge between the observation and the interpretation. If this book has been able to convey such a sense of connection with real physics in nature and to attract the reader to a further study on the physics of plasmas, the principal aims of this book have been achieved.

# Appendix 1

## The Delta Function

The function $\delta(x)$ satisfying

$$\delta(x) = \begin{cases} 0 & (x \neq 0) \\ \infty & (x = 0) \end{cases} \qquad \text{(A1.1)}$$

$$\int_{-\infty}^{\infty} dx\ \delta(x) = 1 \qquad \text{(A1.2)}$$

is called the delta function. For an arbitrary function $f(x)$ defined over the domain $-\infty < x < \infty$, the following relations hold:

$$\int_{-\infty}^{\infty} dx\ f(x)\delta(x) = f(0) \quad , \qquad \text{(A1.3)}$$

$$\int_{a}^{b} dx\ f(x)\delta(x - x_0) = f(x_0) \quad (a < x_0 < b) \quad . \qquad \text{(A1.4)}$$

Various limiting forms exist as analytic expressions for the delta function. For example

$$\delta(x) = \lim_{\eta\to 0} \frac{1}{\eta\sqrt{\pi}} \exp\left(-\frac{x^2}{\eta^2}\right) \qquad \text{(A1.5)}$$

$$\delta(x) = \lim_{\eta\to 0} \frac{\eta}{\pi(x^2 + \eta^2)} = -\lim_{\eta\to 0} \operatorname{Im} \frac{1}{\pi(x + i\eta)} \qquad \text{(A1.6)}$$

satisfy the foregoing relations (Im means the imaginary part).

A three-dimensional delta function $\delta(\vec{\mathbf{r}})$ with respect to the spatial coordinates $\vec{\mathbf{r}} = (x, y, z)$ can be produced from the product of the one-dimensional delta functions as

$$\delta(\vec{\mathbf{r}}) = \delta(x)\delta(y)\delta(z) \quad . \qquad \text{(A1.7)}$$

## Appendix 2

## Fourier Analysis and Application

In this book we frequently use Fourier expansion of physical quantities to describe such effects as fluctuations. This appendix summarizes essential features of the Fourier analysis and some of its applications.

Let a quantity $a(\vec{\mathbf{r}}, t)$ be a function of the space-time variable $\vec{\mathbf{r}}t$, which satisfies the periodic boundary conditions

$$a(\vec{\mathbf{r}}, t) = a(\vec{\mathbf{r}} + n_x L\hat{\mathbf{x}} + n_y L\hat{\mathbf{y}} + n_z L\hat{\mathbf{z}}, t) \quad . \tag{A2.1}$$

Here $\hat{\mathbf{x}}$, $\hat{\mathbf{y}}$, and $\hat{\mathbf{z}}$ are the unit vectors in the x, y and z directions; $n_x$, $n_y$, and $n_z$ represent positive and negative integers including zero. In the main text we sometimes consider the cases with $L = 1$; such a case is called the **periodic boundary conditions for a unit volume.**

The space-time Fourier components of $a(\vec{\mathbf{r}}, t)$ are calculated according to

$$a_{\vec{\mathbf{k}}}(\omega) = \int_{L^3} d\vec{\mathbf{r}} \int_{-\infty}^{\infty} dt \; a(\vec{\mathbf{r}}, t)\exp[-i(\vec{\mathbf{k}} \cdot \vec{\mathbf{r}} - \omega t)] \quad . \tag{A2.2}$$

Since we have adopted the periodic boundary conditions for a cube of volume $L^3$, the wave vector $\vec{k}$ takes on discrete values

$$\vec{k} = \frac{2\pi}{L}(n_x, n_y, n_z) \quad . \tag{A2.3}$$

The original function $a(\vec{r}, t)$ is expressed in terms of the Fourier components (A2.2) via the inverse Fourier transformation. This relation is given by

$$a(\vec{r}, t) = \frac{1}{L^3}\sum_{\vec{k}} \int_{-\infty}^{\infty} \frac{d\omega}{2\pi}\, a_{\vec{k}}(\omega)\exp[i(\vec{k}\cdot\vec{r} - \omega t)] \tag{A2.4}$$

where the summation with respect to $\vec{k}$ is carried out over those discrete values specified by Eq. (A2.3).

As Eq. (A2.3) indicates, a volume $(2\pi/L)^3$ in the three-dimensional wave-number space $(k_x, k_y, k_z)$ corresponds to a wave vector pertaining to the summation of Eq. (A2.4). In the limit of $L \to \infty$, the summation in Eq. (A2.4) turns into a three-dimensional integral with respect to $\vec{k}$. In this case, since

$$\sum_{\vec{k}} \to L^3 \int_{-\infty}^{\infty} \frac{dk_x}{2\pi} \int_{-\infty}^{\infty} \frac{dk_y}{2\pi} \int_{-\infty}^{\infty} \frac{dk_z}{2\pi} \equiv L^3 \int_{-\infty}^{\infty} \frac{d\vec{k}}{(2\pi)^3} \quad , \tag{A2.5}$$

Eq. (A2.4) is transformed as

$$a(\vec{r}, t) = \int_{-\infty}^{\infty} \frac{d\vec{k}}{(2\pi)^3} \int_{-\infty}^{\infty} \frac{d\omega}{2\pi}\, a_{\vec{k}}(\omega)\exp[i(\vec{k}\cdot\vec{r} - \omega t)] \quad . \tag{A2.6}$$

The three-dimensional delta function $\delta(\vec{r})$ introduced in Eq. (A1.7) of Appendix 1 is Fourier transformed, by virtue of Eq. (A1.2) or (2.25), as

$$\int d\vec{\mathbf{r}}\ \delta(\vec{\mathbf{r}})\exp(-i\vec{\mathbf{k}}\cdot\vec{\mathbf{r}}) = 1 \quad . \tag{A2.7}$$

Hence we find

$$\delta(\vec{\mathbf{r}}) = \int_{-\infty}^{\infty} \frac{d\vec{\mathbf{k}}}{(2\pi)^3} \exp(i\vec{\mathbf{k}}\cdot\vec{\mathbf{r}}) \tag{A2.8}$$

with the aid of Eq. (A2.6).

As an application of the technique of Fourier analysis, we take up the problem of solving Eq. (2.8) in Section 2.2. First we multiply both sides of Eq. (2.8) by $\exp(-i\vec{\mathbf{k}}\cdot\vec{\mathbf{r}})$ and integrate the result with respect to $\vec{\mathbf{r}}$. Assuming that $\phi(\vec{\mathbf{r}})$ approaches zero sufficiently fast in the limit $r \to \infty$, we obtain by partial integrations

$$\begin{aligned}\int d\vec{\mathbf{r}}\ \exp(-i\vec{\mathbf{k}}\cdot\vec{\mathbf{r}})\nabla^2\phi(\vec{\mathbf{r}}) &= \int d\vec{\mathbf{r}}(i\vec{\mathbf{k}})\exp(-i\vec{\mathbf{k}}\cdot\vec{\mathbf{r}})\cdot\nabla\phi(\vec{\mathbf{r}}) \\ &= -k^2\int d\vec{\mathbf{r}}\ \exp(-i\vec{\mathbf{k}}\cdot\vec{\mathbf{r}})\phi(\vec{\mathbf{r}}) \\ &\equiv -k^2\phi_{\vec{\mathbf{k}}} \quad .\end{aligned} \tag{A2.9}$$

That the assumption mentioned above is satisfied can be confirmed a posteriori. The final line of Eq. (A2.9) is the definition of $\phi_{\vec{\mathbf{k}}}$.

With the aid of Eqs. (A2.7) and (A2.9), we find that Eq. (2.8) reduces to

$$\phi_{\vec{\mathbf{k}}} = \frac{4\pi Z_0 e}{k^2 + \lambda_D^{-2}} \quad . \tag{A2.10}$$

Carrying out the inverse Fourier transformation of Eq. (A2.10) in accord with Eq. (A2.6), where $\theta$ refers to the angle between

$\vec{k}$ and $\vec{r}$, we obtain

$$\phi(\vec{r}) = \frac{1}{(2\pi)^3} \int_0^\pi 2\pi \sin\theta d\theta \int_0^\infty k^2 dk \frac{4\pi Z_0 e}{k^2 + \lambda_D^{-2}} \exp(-ik\ r \cos\theta)$$

$$= \frac{2Z_0 e}{\pi r} \int_0^\infty dk \frac{k}{k^2 + \lambda_D^{-2}} \sin(kr) \quad .$$

We note that the integrand is an even function of k and that sin(kr) is the imaginary part (Im) of exp(ikr) = cos(kr) + isin(kr). Hence we may rewrite the equation above in the form,

$$\phi(\vec{r}) = \frac{Z_0 e}{\pi r} \operatorname{Im} \int_{-\infty}^\infty dk \frac{k}{k^2 + \lambda_D^{-2}} \exp(ikr) \quad . \qquad \text{(A2.11)}$$

To carry out this integral, we close the contour of integration by an infinite semicircle $C_\infty$ on the upper half of the complex k plane as shown in Fig. A2.1. The integrand of Eq. (A2.11) has the first-order poles at $k = \pm i/\lambda_D$; the residue at $k = i/\lambda_D$ enclosed by the contour of integration is

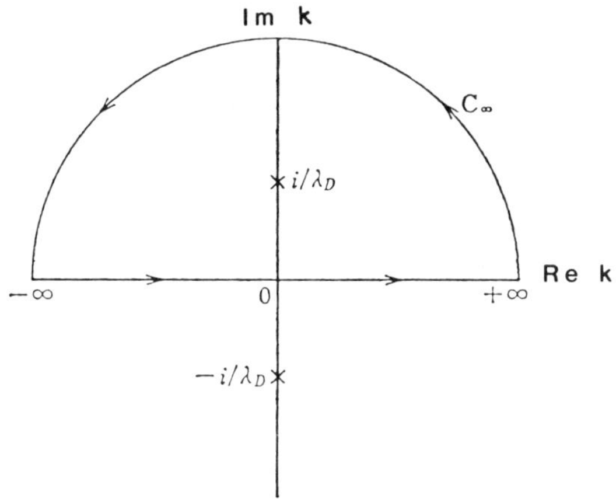


**Figure A2.1.** Contour of integration in Eq. (A2.12).

$$\lim_{k \to i/\lambda_D} \frac{(k - i/\lambda_D)k}{k^2 + \lambda_D^{-2}} \exp(ikr) = \frac{1}{2} \exp(-\frac{r}{\lambda_D}) \quad .$$

According to Cauchy's theorem, we obtain

$$\int_{-\infty}^{\infty} dk \frac{k}{k^2 + \lambda_D^{-2}} \exp(ikr) + \int_{C_\infty} dk \frac{k}{k^2 + \lambda_D^{-2}} \exp(ikr)$$

$$= i\pi \exp(-\frac{r}{\lambda_D}) \quad . \tag{A2.12}$$

The second term on the left-hand side of Eq. (A2.12) is the integration along the infinite semicircle on the upper half plane shown in Fig. A2.1. We transform it into an integration from 0 to $\pi$ with respect to $\theta$ by putting $k = K \exp(i\theta)$; in the limit of $K \to \infty$ this term vanishes. Substitution of Eq. (A2.12) into Eq. (A2.11) yields Eq. (A2.9).

If in particular we let the second term on the left-hand side of Eq. (2.8) approach 0, i.e., $\lambda_D^{-2} \to 0$, $\phi(\vec{r})$ should coincide with $\phi_0(\vec{r})$ of Eq. (2.4). Consequently the Fourier transformation of the ordinary Coulomb potential, Eq. (2.4), is given by Eq. (A2.10) where $\lambda_D^{-2} \to 0$ is taken, that is,

$$\frac{Ze}{r} = \frac{1}{L^3} \sum_{\vec{k}} \frac{4\pi Ze}{k^2} \exp(i\vec{k} \cdot \vec{r}) \quad . \tag{A2.13}$$

## Appendix 3

## Plasma Kinetic Equation and Fluctuations

We consider a classical plasma in a box of volume V consisting of a mixture of charged particles of species "$\sigma$," each of which has an electric charge $e_\sigma$, a mass $m_\sigma$, and the average number density $n_\sigma = N_\sigma/V$. The condition for charge neutrality, $\Sigma_\sigma e_\sigma n_\sigma = 0$, is assumed.

The Klimontovich distribution (Klimontovich 1957; Ichimaru 1973)

$$N_\sigma(X;\ t) = \sum_{j=1}^{N_\sigma} \delta[X - X_j(t)] \tag{A3.1}$$

describes the microscopic density distribution for the particles of the $\sigma$ species in the 6-dimensional phase space,

$$X = (\vec{r},\ \vec{p}) \quad , \tag{A3.2}$$

and $X_j(t)$ represents the trajectory of the jth particle in the phase space.

The distribution (A3.1) satisfies the Klimontovich equation

$$\frac{\partial N_\sigma}{\partial t} + \vec{\mathbf{v}} \cdot \frac{\partial N_\sigma}{\partial \vec{\mathbf{r}}} + \dot{\vec{\mathbf{p}}} \cdot \frac{\partial N_\sigma}{\partial \vec{\mathbf{p}}} = 0 \tag{A3.3}$$

where

$$\vec{\mathbf{p}} = \gamma m_\sigma \vec{\mathbf{v}}$$

$$\varepsilon = (\gamma - 1) m_\sigma c^2$$

$$\gamma = (1 - v^2/c^2)^{-1/2} \quad . \tag{A3.4}$$

Those are the momentum, the energy (minus the rest-mass energy), and the Lorentz factor of a single particle. The force $\dot{\vec{\mathbf{p}}}$ in Eq. (A3.3) is given by

$$\dot{\vec{\mathbf{p}}} = e_\sigma [\vec{\mathbf{E}} + \frac{\vec{\mathbf{v}}}{c} \times \vec{\mathbf{B}}] \quad . \tag{A3.5}$$

This is the Lorentz force acting on a particle at $X = (\vec{\mathbf{r}}, \vec{\mathbf{p}})$ due to **microscopic** (and external) electromagnetic fields, $\vec{\mathbf{E}}$ and $\vec{\mathbf{B}}$.

The electromagnetic fields are determined from the solution to the set of Maxwell equations:

$$\nabla \times \vec{\mathbf{E}} + \frac{1}{c} \frac{\partial \vec{\mathbf{B}}}{\partial t} = 0 \tag{A3.6a}$$

$$\nabla \times \vec{\mathbf{B}} - \frac{1}{c} \frac{\partial \vec{\mathbf{E}}}{\partial t} = \frac{4\pi}{c} \vec{\mathbf{J}} \tag{A3.6b}$$

$$\nabla \cdot \vec{E} = 4\pi\rho_e \tag{A3.6c}$$

$$\nabla \cdot \vec{B} = 0 \quad , \tag{A3.6d}$$

where the electric current density $\vec{J}$ and the electric charge density $\rho_e$ are given by the sums of the microscopic and external contributions:

$$\vec{J} = \sum_\sigma e_\sigma \int d\vec{\mathbf{p}}\, \vec{\mathbf{v}}\, N_\sigma(X;\ t) + \vec{J}_{ext} \quad , \tag{A3.7a}$$

$$\rho_e = \sum_\sigma e_\sigma \int d\vec{\mathbf{p}}\, N_\sigma(X;\ t) + \rho_{ext} \quad . \tag{A3.7b}$$

The fine-grained quantities such as Eqs. (A3.1) and (A3.7), although precise in describing microscopic states of the many-particle system, would not by themselves correspond to the coarse-grained quantities which we observe in the macroscopic world. To establish a connection between the microscopic and macroscopic descriptions, we introduce an averaging process over the ensemble of systems with the same "macroscopic realizations" as those of the plasma system under consideration, which we shall denote by angular brackets. In practice we may reinterpret it as representing a space-time average of the microscopically fluctuating quantities over scales substantially greater than those associated with the microscopic fluctuations.

We may thus split the Klimontovich distribution (A3.1) into its average and fluctuations:

$$N_\sigma(X;\ t) = \langle N_\sigma(X;\ t)\rangle + \delta N_\sigma(X;\ t) \quad . \tag{A3.8}$$

The average part then represents the single-particle distribution function,

$$\langle N_\sigma(X;\, t)\rangle = F_\sigma(X;\, t) \quad , \tag{A3.9}$$

with normalization

$$\int d\vec{\mathbf{p}}\, F_\sigma(X;\, t) = n_\sigma(\vec{\mathbf{r}},\, t) \quad , \tag{A3.10}$$

the (macroscopic) number density of the particles of σ species. For a homogeneous and stationary system we write $F_\sigma(X;\, t) = F_\sigma(\vec{\mathbf{p}})$ and $n_\sigma(\vec{\mathbf{r}},\, t) = n_\sigma$.

The **kinetic equations** are those which govern the evolution of the single-particle distribution functions in the phase space. Taking the average of Eq. (A3.3), we obtain

$$\frac{\partial F_\sigma}{\partial t} + \vec{v} \cdot \frac{\partial F_\sigma}{\partial \vec{r}} + e_\sigma\left[\langle\vec{\mathbf{E}}\rangle + \frac{\vec{\mathbf{v}}}{c} \times \langle\vec{\mathbf{B}}\rangle\right] \cdot \frac{\partial F_\sigma}{\partial \vec{\mathbf{p}}}$$

$$= -e_\sigma \frac{\partial}{\partial \vec{\mathbf{p}}} \cdot \langle \delta N_\sigma\left(\delta\vec{\mathbf{E}} + \frac{\vec{\mathbf{v}}}{c} \times \delta\vec{\mathbf{B}}\right)\rangle \quad , \tag{A3.11}$$

where we have expressed the electromagnetic fields as sums of the average parts and the fluctuations:

$$\vec{\mathbf{E}} = \langle\vec{\mathbf{E}}\rangle + \delta\vec{\mathbf{E}} \quad , \tag{A3.12}$$

$$\vec{\mathbf{B}} = \langle\vec{\mathbf{B}}\rangle + \delta\vec{\mathbf{B}} \quad .$$

The right-hand side of Eq. (A3.11), which we shall denote as $\partial F_\sigma/\partial t]_c$, represents the **collision term** of the kinetic equation. By its construction the collision term satisfies the **conservation laws**:

$$\int d\vec{\mathbf{p}} \left.\frac{\partial F_\sigma}{\partial t}\right]_c = 0 \quad , \tag{A3.13a}$$

$$\int d\vec{\mathbf{p}}\ \vec{\mathbf{p}} \left.\frac{\partial F_\sigma}{\partial t}\right]_c = \langle \delta\rho_{e\sigma} \delta\vec{\mathbf{E}} \rangle + \frac{1}{c} \langle \delta\vec{\mathbf{J}}_\sigma \times \delta\vec{\mathbf{B}} \rangle \quad , \tag{A3.13b}$$

$$\int d\vec{\mathbf{p}}\ \varepsilon \left.\frac{\partial F_\sigma}{\partial t}\right]_c = \langle \delta\vec{\mathbf{J}}_\sigma \cdot \delta\vec{\mathbf{E}} \rangle \quad , \tag{A3.13c}$$

where

$$\delta\vec{\mathbf{J}}_\sigma = e_\sigma \int d\vec{p}\ \vec{\mathbf{v}} \delta N_\sigma(X;\ t) \tag{A3.14a}$$

$$\delta\rho_{e\sigma} = e_\sigma \int d\vec{\mathbf{p}}\ \delta N_\sigma(X;\ t) \quad . \tag{A3.14b}$$

Subtraction of Eq. (A3.11) from Eq. (A3.3) yields the equations for the fluctuating parts, $\delta N_\sigma(X;\ t)$, of the Klimontovich distribution:

$$\frac{\partial \delta N_\sigma}{\partial t} + \vec{\mathbf{v}} \cdot \frac{\partial \delta N_\sigma}{\partial \vec{\mathbf{r}}} + e_\sigma \left[ \langle \vec{\mathbf{E}} \rangle + \frac{\vec{\mathbf{v}}}{c} \times \langle \vec{\mathbf{B}} \rangle \right] \cdot \frac{\partial \delta N_\sigma}{\partial \vec{p}}$$

$$= - e_\sigma \left[ \delta\vec{\mathbf{E}} + \frac{\vec{\mathbf{v}}}{c} \times \delta\vec{\mathbf{B}} \right] \cdot \frac{\partial F_\sigma}{\partial \vec{\mathbf{p}}} - e_\sigma \left[ \delta\vec{\mathbf{E}} + \frac{\vec{\mathbf{v}}}{c} \times \delta\vec{\mathbf{B}} \right] \cdot \frac{\partial \delta N_\sigma}{\partial \vec{\mathbf{p}}}$$

$$+ e_\sigma \left\langle \left[ \delta\vec{\mathbf{E}} + \frac{\vec{\mathbf{v}}}{c} \times \delta\vec{\mathbf{B}} \right] \cdot \frac{\partial \delta N_\sigma}{\partial \vec{\mathbf{p}}} \right\rangle \quad . \tag{A3.15}$$

For brevity we write this as

$$\frac{\partial}{\partial t} \delta N_\sigma(X;\ t) + \mathscr{L}(X;\ t) \delta N_\sigma(X;\ t) = S(X;\ t) \quad , \tag{A3.16}$$

where

$$\mathcal{L}(X;\ t) = \vec{v} \cdot \frac{\partial}{\partial \vec{r}} + e_\sigma[\langle\vec{E}\rangle + \frac{\vec{v}}{c} \times \langle\vec{B}\rangle] \cdot \frac{\partial}{\partial \vec{p}} \qquad \text{(A3.17)}$$

is a differential operator in the phase space, and $S(X;\ t)$ denotes the entirety of the right-hand side of Eq. (A3.15).

The form of Eq. (A3.16) suggests that $S(X;\ t)$, being the inhomogeneous term of the equation, plays the part of the source term for the evolution of $\delta N_\sigma(X;\ t)$. We thus seek for the solution to Eq. (A3.16) in a form

$$\delta N_\sigma(X;\ t) = \delta N_\sigma^{(0)}(X;\ t) + \delta N_\sigma^{(s)}(X;\ t) \quad , \qquad \text{(A3.18)}$$

where $\delta N_\sigma^{(s)}(X;\ t)$ is a special solution to Eq. (A3.16) and $\delta N_\sigma^{(0)}(X;\ t)$ represents a general solution to the homogeneous equation:

$$\frac{\partial}{\partial t}\,\delta N_\sigma^{(0)}(X;\ t) + \mathcal{L}(X;\ t)\,\delta N_\sigma^{(0)}(X;\ t) = 0 \quad . \qquad \text{(A3.19)}$$

As the ensuing considerations will clarify, $\delta N_\sigma^{(0)}$ corresponds to those **spontaneous** fluctuations arising from discreteness of the particles which exist even in the absence of particle interactions, while $\delta N_\sigma^{(s)}$ describes those **induced** fluctuations through the microscopic electromagnetic interactions between discrete particles.

Special solution to Eq. (A3.16) can be obtained by following the method of integration along the unperturbed trajectories of the particles in the phase space. Assuming that the macroscopic electromagnetic fields, $\langle\vec{E}\rangle$ and $\langle\vec{B}\rangle$, are known functions in space and time, we set equations of motion for a particle in the average fields,

$$\frac{d\vec{r}}{dt} = \vec{v} \quad , \tag{A3.20a}$$

$$\frac{d\vec{p}}{dt} = e_\sigma\left[\langle\vec{E}\rangle + \frac{\vec{v}}{c} \times \langle\vec{B}\rangle\right] \quad . \tag{A3.20b}$$

The solution to this set of equations determines the unperturbed phase-space trajectory, $X' = (\vec{r}', \vec{p}')$ at $t'$, when the "initial" condition is given as $X = (\vec{r}, \vec{p})$ at $t$. We call it "unperturbed" because Eq. (A3.20b) does not contain the effects of the fluctuating electromagnetic fields.

We can thus invert the differential operators appearing on the left-hand side of Eq. (A3.16) by integration along those unperturbed trajectories, to find

$$\delta N_\sigma^{(s)}(X;\ t) = \int_{-\infty}^{t} dt'\ S(X';\ t') \quad . \tag{A3.21}$$

This solution incorporates the physical principle of causality by setting the lower limit of the integration to be minus infinity; the fluctuations are built up through adiabatic turning on of the electromagnetic interaction from the infinite past of time.

To obtain $\delta N_\sigma^{(0)}(X;\ t)$, we find it instructive to note that Eq. (A3.19) may be derived by putting $\delta\vec{E} = 0$ and $\delta\vec{B} = 0$ in Eq. (A3.15). Since the microscopic interactions between the discrete particles are represented by $\delta\vec{E}$ and $\delta\vec{B}$, we interpret that Eq. (A3.19) describes the evolution in the phase space of the fluctuations in a fictitious system in which the microscopic interactions are suppressed; $\delta N_\sigma^{(0)}(X;\ t)$ thus refers to those fluctuations in an "ideal-gas" system consisting of discrete particles,

$$\delta N_\sigma^{(0)}(X;\ t) = \sum_{j=1}^{N_\sigma} \delta[X - X_j^{(0)}(t,\ t_0)]$$

$$- < \sum_{j=1}^{N_\sigma} \delta[X - X_j^{(0)}(t,\ t_0)]> \quad . \qquad (A3.22)$$

Here

$$X_j^{(0)}(t,\ t_0) = [\vec{r}_j(t,\ t_0),\ \vec{p}_j(t,\ t_0)] \qquad (A3.23)$$

represents the unperturbed phase-space trajectory of the jth particle with the initial condition $[\vec{r}_j(t_0),\ \vec{p}_j(t_0)]$ at $t_0$.

# Appendix 4

## Dielectric Tensor

Let us consider the evolution of the fluctuations, treated generally in the previous appendix, for a more specific case in which the plasma may be regarded as **uniform** and **stationary** over scales greater than those associated with the microscopic fluctuations. As far as the fluctuation analysis is concerned, we thus assume the system to be translationally invariant both in space and time. In these circumstances it is convenient to work with the Fourier components (cf. Appendix 2) of the fluctuating quantities, so that for example

$$\delta\tilde{N}_\sigma(\vec{\mathbf{k}}, \omega; \vec{\mathbf{p}}) = \int_V d\vec{\mathbf{r}} \int_{-\infty}^{\infty} dt\, \delta N_\sigma(X; t) \exp[-i(\vec{\mathbf{k}} \cdot \vec{\mathbf{r}} - \omega t)] \quad , \tag{A4.1}$$

where V refers to the plasma volume under consideration. The inverse transformation is

$$\delta N_\sigma(X; t) = \frac{1}{V} \sum_{\vec{\mathbf{k}}} \int_{-\infty}^{\infty} \frac{d\omega}{2\pi}\, \delta\tilde{N}_\sigma(\vec{\mathbf{k}}, \omega; \vec{\mathbf{p}}) \exp[i(\vec{\mathbf{k}} \cdot \vec{\mathbf{r}} - \omega t)] \quad . \tag{A4.2}$$

Other fluctuating quantities can be analogously expressed.

Substituting Eq. (A3.21) in Eq. (A3.18) with the aid of Eqs. (A3.15) and (A3.16), we find

$$\delta\tilde{N}_\sigma(\vec{k},\ \omega;\ \vec{p}) = \delta\tilde{N}_\sigma^{(0)}(\vec{k},\ \omega;\ \vec{p})$$

$$- e_\sigma \int_0^\infty d\tau \left[\frac{\partial F_\sigma(\vec{p}\,')}{\partial \vec{p}\,'}\right] \cdot \overleftrightarrow{T}(\vec{k},\ \omega;\ \vec{p}\,') \cdot \delta\vec{E}(\vec{k},\ \omega) \exp[-i\phi(\tau) - \eta\tau]$$

$$- \frac{e}{V} \sum_{\vec{q}} \int_{-\infty}^{\infty} \frac{d\omega'}{2\pi} \int_0^\infty d\tau \left[\frac{\partial}{\partial \vec{p}\,'} \delta\tilde{N}_\sigma(\vec{k} - \vec{q},\ \omega - \omega';\ \vec{p}\,')\right] \cdot \overleftrightarrow{T}(\vec{q},\ \omega';\ \vec{p}\,')$$

$$\cdot\ \delta\vec{E}(\vec{q},\ \omega') \exp[-1\phi(\tau) - \eta\tau] \qquad \text{(A4.3)}$$

where $\tau = t - t'$

$$\phi(\tau) = \vec{k} \cdot (\vec{r} - \vec{r}\,') - \omega\tau \qquad \text{(A4.4)}$$

$$\overleftrightarrow{T}(\vec{k},\ \omega;\ \vec{p}) = \left(1 - \frac{\vec{k} \cdot \vec{v}}{\omega}\right)\overleftrightarrow{I} + \frac{\vec{k}\,\vec{v}}{\omega}\ , \qquad \text{(A4.5)}$$

and

$$\overleftrightarrow{I} = \begin{pmatrix} 1 & 0 & 0 \\ 0 & 1 & 0 \\ 0 & 0 & 1 \end{pmatrix} . \qquad \text{(A4.6)}$$

In the derivation of Eq. (A4.3), the magnetic field fluctuation $\delta\vec{\mathbf{B}}(\vec{\mathbf{k}}, \omega)$ has been expressed in terms of the electric field fluctuation via Eq. (A3.6a); the tensor $\overleftrightarrow{T}(\vec{\mathbf{k}}, \omega; \vec{\mathbf{p}})$ accounts for such a connection. The use of a **positive infinitesimal** $\eta$ in Eq. (A4.3) is related to the causal boundary conditions adopted in Eq. (A3.21). At $t' = -\infty$ (or $\tau = \infty$) the fluctuations induced by the electromagnetic interactions (i.e., $\delta\vec{\mathbf{E}}$ and $\delta\vec{\mathbf{B}}$) did not exist; the interactions are adiabatically (i.e., $\eta = +0$) turned on to produce the induced fluctuations with the present strength at $t' = t$ (or $\tau = 0$).

The first term on the right-hand side of Eq. (A4.3) represents the spontaneous fluctuations arising from the particle discreteness in the fictitious "ideal-gas" counterpart. The second term corresponds to the fluctuations induced **linearly** with respect to the electromagnetic fluctuations. The third term describes those **nonlinear** fluctuations arising from the mode coupling between the density fluctuations $\delta\tilde{N}_\sigma(\vec{\mathbf{k}} - \vec{\mathbf{q}}, \omega - \omega'; \vec{\mathbf{p}}')$ and the electromagnetic fluctuations $\delta\vec{\mathbf{E}}(\vec{\mathbf{q}}, \omega')$.

As explained in Section 2.6, the effects arising from the nonlinear coupling between fluctuations are neglected in the random-phase approximation (RPA). The electric-current density induced linearly by the electromagnetic fluctuations is then calculated as

$$\delta\vec{\mathbf{J}}^{\text{ind}}(\vec{\mathbf{k}}, \omega) = \sum_\sigma e_\sigma \int d\vec{\mathbf{p}}\, \vec{\mathbf{v}}\,\delta\tilde{N}_\sigma^{(s)}(\vec{\mathbf{k}}, \omega; \vec{\mathbf{p}})$$

$$= - \sum_\sigma e_\sigma^2 \int d\vec{\mathbf{p}} \int_0^\infty d\tau\, \vec{\mathbf{v}}\left[\frac{\partial F_\sigma(\vec{\mathbf{p}}')}{\partial \vec{p}'}\right] \cdot \overleftrightarrow{T}(\vec{\mathbf{k}}, \omega; \vec{\mathbf{p}}')$$

$$\cdot\ \delta\vec{E}(\vec{\mathbf{k}}, \omega)\ \exp[-i\phi(\tau) - \eta\tau] \quad . \tag{A4.7}$$

The frequency and wave-vector dependent **electric conductivity tensor** $\overleftrightarrow{\sigma}(\vec{\mathbf{k}}, \omega)$ and the **dielectric tensor** $\overleftrightarrow{\varepsilon}(\vec{\mathbf{k}}, \omega)$ may then be introduced by the equation

$$\delta\vec{\mathbf{J}}^{ind}(\vec{\mathbf{k}}, \omega) = \overleftrightarrow{\sigma}(\vec{\mathbf{k}}, \omega) \cdot \delta\vec{\mathbf{E}}(\vec{\mathbf{k}}, \omega)$$

$$= -\frac{i\omega}{4\pi}[\overleftrightarrow{\varepsilon}(\vec{\mathbf{k}}, \omega) - \overleftrightarrow{I}] \cdot \delta\vec{\mathbf{E}}(\vec{\mathbf{k}}, \omega) \quad . \tag{A4.8}$$

Comparison between Eqs. (A4.7) and (A4.8) yields after partial integrations

$$\overleftrightarrow{\varepsilon}(\vec{\mathbf{k}}, \omega) = \overleftrightarrow{I} - \sum_\sigma \frac{4\pi e_\sigma^2}{\omega^2} \int d\vec{\mathbf{p}}\left(\frac{\partial\vec{\mathbf{v}}}{\partial\vec{\mathbf{p}}}\right)F_\sigma(\vec{\mathbf{p}})$$

$$+ \sum_\sigma \frac{4\pi e_\sigma^2}{\omega^2} \int d\vec{\mathbf{p}} \int_0^\infty d\tau\, \vec{\mathbf{v}}\left[\left(\frac{\partial}{\partial\tau} - i\vec{\mathbf{v}}'\vec{\mathbf{k}}\cdot\right) \frac{\partial F_\sigma(\vec{\mathbf{p}}')}{\partial\vec{\mathbf{p}}'}\right]$$

$$\times \exp[-i\phi(\tau) - \eta\tau] \quad . \tag{A4.9}$$

The tensor $\partial\vec{\mathbf{v}}/\partial\vec{\mathbf{p}}$ represents the **inverse mass** tensor, which is isotropic for a nonrelativistic plasma.

Further evaluation of Eq. (A4.9) depends on specification of the unperturbed trajectory described by Eqs. (A3.20). When $\langle\vec{\mathbf{E}}\rangle = 0$ and $\langle\vec{\mathbf{B}}\rangle = 0$, we have $\vec{\mathbf{r}} - \vec{\mathbf{r}}' = \vec{\mathbf{v}}\tau$ and $\vec{\mathbf{p}} = \vec{\mathbf{p}}'$, so that

$$\overleftrightarrow{\varepsilon}(\vec{\mathbf{k}}, \omega) = \overleftrightarrow{I} - \sum_\sigma \frac{4\pi e_\sigma^2}{\omega^2} \int d\vec{\mathbf{p}}\left(\frac{\partial\vec{\mathbf{v}}}{\partial\vec{\mathbf{p}}}\right)F_\sigma(\vec{\mathbf{p}})$$

$$- \sum_\sigma \frac{4\pi e_\sigma^2}{\omega^2} \int d\vec{\mathbf{p}} \frac{\vec{\mathbf{v}}\,\vec{\mathbf{v}}}{\vec{\mathbf{k}}\cdot\vec{\mathbf{v}} - \omega - i\eta}\, \vec{\mathbf{k}} \cdot \frac{\partial F_\sigma(\vec{\mathbf{p}})}{\partial\vec{\mathbf{p}}} \quad . \tag{A4.10}$$

When $\langle\vec{E}\rangle = 0$ and $\langle\vec{B}\rangle = B\hat{z}$, we find after lengthy calculations (see e.g., Ichimaru 1973)

$$\overleftrightarrow{\varepsilon}(\vec{k}, \omega) = \overleftrightarrow{I} - \sum_\sigma \frac{4\pi e_\sigma^2}{\omega^2} \int d\vec{p}\left(\frac{\partial\vec{v}}{\partial\vec{p}}\right)F_\sigma(\vec{p})$$

$$- \sum_\sigma \frac{4\pi e_\sigma^2}{\omega^2} \sum_{\nu=-\infty}^{\infty} \int d\vec{p}\left[\frac{\nu\Omega_\sigma}{v_\perp}\frac{\partial F_\sigma}{\partial p_\perp} + k_\parallel \frac{\partial F_\sigma}{\partial p_\parallel}\right]$$

$$\times \frac{\overleftrightarrow{\Pi}_\sigma(p_\perp, p_\parallel; \nu)}{\nu\Omega_\sigma + k_\parallel v_\parallel - \omega - i\eta} , \tag{A4.11}$$

$$\overleftrightarrow{\Pi}_\sigma(p_\perp, p_\parallel; \nu) = \begin{pmatrix} \frac{\nu^2\Omega_\sigma^2}{k_\perp^2} J_\nu^2 & iv_\perp \frac{\nu\Omega_\sigma}{k_\perp} J_\nu J_\nu' & v_\parallel \frac{\nu\Omega_\sigma}{k_\perp} J_\nu^2 \\ -iv_\perp \frac{\nu\Omega_\sigma}{k_\perp} J_\nu J_\nu' & v_\perp^2 (J_\nu')^2 & -iv_\parallel v_\perp J_\nu J_\nu' \\ v_\parallel \frac{\nu\Omega_\sigma}{k_\perp} J_\nu^2 & iv_\parallel v_\perp J_\nu J_\nu' & v_\parallel^2 J_\nu^2 \end{pmatrix} \tag{A4.12}$$

where $J_\nu = J_\nu(\zeta)$ and $J_\nu' = dJ_\nu(\zeta)/d\zeta$ are the Bessel function of the $\nu$th order and its derivative with $\vec{k} = k_\perp\hat{x} + k_\parallel\hat{z}$,

$$\Omega_\sigma = e_\sigma B/m_\sigma c\gamma \tag{A4.13}$$

$$\zeta = k_\perp v_\perp/\Omega_\sigma \quad . \tag{A4.14}$$

In Eq. (A4.10), $F_\sigma(\vec{p})$ is a function of $p_\perp$ and $p_\parallel$, which are defined analogously to $v_\perp$ and $v_\parallel$ in Eq. (5.2);

$$\int d\vec{\mathbf{p}} = 2\pi \int_0^\infty p_\perp dp_\perp \int_{-\infty}^\infty dp_\parallel \quad . \tag{A4.15}$$

## Appendix 5

## Dielectric Response Functions in the Random-Phase Approximation

We begin this appendix by considering a degenerate electron plasma at $T = 0$ as treated in Chapter 6. A weak external potential field, Eq. (6.8), is applied to such a system and we aim at calculating a density response, Eq. (6.9), linear to the disturbance.

The Hamiltonian of the system (with a unit volume) is given by

$$H_{tot} = H + H_{ext} \tag{A5.1}$$

$$H = \sum_{\vec{\mathbf{p}}\sigma} \left(\frac{p^2}{2m}\right) c^+_{\vec{\mathbf{p}}\sigma} c_{\vec{\mathbf{p}}\sigma} + \frac{1}{2} \sum_{\vec{\mathbf{q}}}{}' v(q) \left[\rho_{\vec{\mathbf{q}}} \rho_{-\vec{\mathbf{q}}} - n\right] \tag{A5.2}$$

$$H_{ext} = -e(\phi_{ext})_{\vec{\mathbf{k}}\omega} \rho_{-\vec{\mathbf{k}}} \exp(-i\omega t + \eta t)$$

$$+ \text{(hermitian conjugate)} \tag{A5.3}$$

with

$$v(q) = 4\pi e^2/q^2 \tag{A5.4}$$

and the prime on the summation sign in Eq. (A5.2) implies that the term with $\vec{\mathbf{q}} = 0$ is to be omitted in the summation. Here $c^+_{\vec{\mathbf{p}}\sigma}$ and $c_{\vec{\mathbf{p}}\sigma}$ represent the creation and annihilation operators of an electron with the momentum $\vec{\mathbf{p}}$ and spin $\sigma$, which satisfy the anticommutator relations for the fermions (see e.g., Pines 1963):

$$\{c^+_{\vec{\mathbf{p}}\sigma}, c^+_{\vec{\mathbf{p}}'\sigma'}\} = \{c_{\vec{\mathbf{p}}\sigma}, c_{\vec{\mathbf{p}}'\sigma'}\} = 0$$

$$\{c_{\vec{\mathbf{p}}\sigma}, c^+_{\vec{\mathbf{p}}'\sigma'}\} = \delta_{\vec{\mathbf{p}}\vec{\mathbf{p}}'}\delta_{\sigma\sigma'} \quad . \tag{A5.5}$$

In these equations, $\{A, B\} = AB + BA$; $\delta_{\vec{\mathbf{p}}\vec{\mathbf{p}}'}$ and $\delta_{\sigma\sigma'}$ are Kronecker's deltas.

For the calculation of the density response, it is convenient to work with the operator representing an electron-hole pair:

$$\rho_{\vec{\mathbf{p}}\vec{\mathbf{k}}\sigma} = c^+_{\vec{\mathbf{p}}\sigma} c_{\vec{\mathbf{p}} + \hbar\vec{\mathbf{k}}\sigma} \quad . \tag{A5.6}$$

The Fourier components of the density fluctuations are then given by

$$\rho_{\vec{\mathbf{k}}} = \sum_{\vec{\mathbf{p}}\sigma} \rho_{\vec{\mathbf{p}}\vec{\mathbf{k}}\sigma} \quad . \tag{A5.7}$$

The occupation-number operator for the state $(\vec{\mathbf{p}}, \sigma)$ is expressed as

$$n_{\vec{p}\sigma} = \rho_{\vec{p}\vec{k}\sigma}\Big|_{\vec{k}=0} \quad . \tag{A5.8}$$

The Heisenberg equation of motion for $\rho_{\vec{p}\vec{k}\sigma}$ is

$$i\hbar \frac{\partial}{\partial t} \rho_{\vec{p}\vec{k}\sigma} = [\rho_{\vec{p}\vec{k}\sigma}, H_{tot}] \tag{A5.9}$$

where [ , ] denotes a commutator. Explicit calculation with the aid of Eqs. (A5.2) and (A5.3) yields

$$\begin{aligned} i\hbar \frac{\partial}{\partial t} \rho_{\vec{p}\vec{k}\sigma} &= \hbar\omega_{\vec{p}\vec{k}}\rho_{\vec{p}\vec{k}\sigma} \\ &- \frac{1}{2} v(k) \{\rho_{\vec{k}}, \Delta_{\vec{p}}^{\vec{k}} n_{\vec{p}\sigma}\} \\ &+ \frac{1}{2} \sum_{\vec{q}(\neq\vec{k})}{}' v(q) \{\rho_{\vec{q}}, \Delta_{\vec{p}}^{\vec{q}} \rho_{\vec{p}\vec{k}-\vec{q}\sigma}\} \\ &- e(\phi_{ext})_{\vec{k}\omega} \Delta_{\vec{p}}^{\vec{k}} n_{\vec{p}\sigma} \exp(-i\omega t + \eta t) \end{aligned} \tag{A5.10}$$

where

$$\omega_{\vec{p}\vec{k}} = \frac{\vec{k} \cdot \vec{p}}{m} + \frac{\hbar k^2}{2m} \tag{A5.11}$$

and $\Delta_{\vec{p}}^{\vec{q}}$ is a displacement operator defined as

$$\Delta_{\vec{\mathbf{p}}}^{\vec{\mathbf{q}}} \rho_{\vec{\mathbf{p}}\vec{\mathbf{k}}\sigma} = \rho_{\vec{\mathbf{p}}\vec{\mathbf{k}}\sigma} - \rho_{\vec{\mathbf{p}} + \hbar\vec{\mathbf{q}}\vec{\mathbf{k}}\sigma} \quad . \tag{A5.12}$$

The third term on the right-hand side of Eq. (A5.10) describes the effect of nonlinear coupling between density fluctuations. As explained in Section 2.6, such a contribution is neglected in the random-phase approximation (RPA). We then set $\rho_{\vec{\mathbf{p}}\vec{\mathbf{k}}\sigma}$ as a sum of the unperturbed part $\rho_{\vec{\mathbf{p}}\vec{\mathbf{k}}\sigma}^{(0)}$ and the induced perturbation $\delta\rho_{\vec{\mathbf{p}}\vec{\mathbf{k}}\sigma} \exp(-i\omega t + \eta t)$, and linearize Eq. (A5.10) with respect to the perturbation. The induced density fluctuation is finally calculated as

$$\langle\rho_{\text{ind}}\rangle_{\vec{\mathbf{k}}\omega} = \langle \sum_{\vec{\mathbf{p}}\sigma} \delta\rho_{\vec{\mathbf{p}}\vec{\mathbf{k}}\sigma} \rangle \quad . \tag{A5.13}$$

The dielectric response function $\varepsilon(\vec{\mathbf{k}}, \omega)$ is defined by Eq. (6.10) with $Z = -1$. Carrying out the calculation described above, we find the RPA dielectric function or the **Lindhard dielectric function** (Lindhard 1954),

$$\varepsilon(\vec{\mathbf{k}}, \omega) = 1 - v(k)\chi_0(\vec{\mathbf{k}}, \omega) \ , \tag{A5.14}$$

where

$$\chi_0(\vec{\mathbf{k}}, \omega) = \frac{2}{\hbar} \sum_{\vec{\mathbf{p}}} \frac{1}{\omega - \omega_{\vec{\mathbf{p}}\vec{\mathbf{k}}} + i\eta} \Delta_{\vec{\mathbf{p}}}^{\vec{\mathbf{k}}} F(\vec{\mathbf{p}}) \tag{A5.15}$$

is the **free-electron polarizability** with the Fermi distribution $F(\vec{\mathbf{p}}) = \langle n_{\vec{\mathbf{p}}\sigma}\rangle$ given by Eq. (6.18).

The static dielectric function is obtained by setting $\omega = 0$ in Eq. (A5.14). We find

$$\varepsilon(\vec{k}, 0) = 1 + \frac{k_{TF}^2}{k^2} \left\{\frac{1}{2} + \frac{k_F}{2k}\left(1 - \frac{k^2}{4k_F^2}\right)\ln\left|\frac{k + 2k_F}{k - 2k_F}\right|\right\} \tag{A5.16}$$

where

$$k_{TF} = \left(\frac{6\pi ne^2}{E_F}\right)^{1/2} = 0.814\ r_s^{1/2} k_F \tag{A5.17}$$

is the **Thomas-Fermi screening constant.**

The RPA dielectric function for a plasma obeying the classical statistics may be derived by taking the classical limit of Eq. (A5.15). With the velocity distribution $f(\vec{v})$ normalized to unity [cf. Eq. (1.2)], we take the classical limit via

$$2\Delta_{\vec{p}}^{\vec{k}} F(\vec{p}) \longrightarrow -\frac{n}{m}\left(\frac{2\pi\hbar}{m}\right)^3 \hbar\vec{k} \cdot \frac{\partial f(\vec{v})}{\partial \vec{v}} \quad , \tag{A5.18a}$$

$$\sum_{\vec{p}} \longrightarrow \left(\frac{m}{2\pi\hbar}\right)^3 \int d\vec{v} \quad , \tag{A5.18b}$$

$$\omega_{\vec{p}\vec{k}} \longrightarrow \vec{k} \cdot \vec{v} \quad , \tag{A5.18c}$$

so that the dielectric function for a classical, nonrelativistic electron plasma takes the form,

$$\varepsilon(\vec{k}, \omega) = 1 + \frac{\omega_p^2}{k^2} \int d\vec{v} \frac{1}{\omega - \vec{k} \cdot \vec{v} + i\eta} \vec{k} \cdot \frac{\partial f(\vec{v})}{\partial \vec{v}} \quad . \tag{A5.19}$$

This is the **Vlasov dielectric function.**

The dispersion relation for the plasma oscillation is given by the solution to Eq. (6.15). Assuming $|\omega/k| \gg \sqrt{\langle v^2 \rangle}$, we expand the factor $(\omega - \vec{k} \cdot \vec{v} + i\eta)^{-1}$ in Eq. (A5.19) in a power series of $\omega^{-1}$. Carrying out the partial integrations and assuming

$$\langle \vec{v} \rangle = \int d\vec{v}\, \vec{v} f(\vec{v}) = 0$$

for a reflectionally invariant system, we find

$$\mathrm{Re}\ \varepsilon(\vec{k}, \omega) = 1 - \frac{\omega_p^2}{\omega^2} - \frac{3\omega_p^2}{\omega^4} \int d\vec{v} (\vec{k} \cdot \vec{v})^2 f(\vec{v}) \ \ldots\ldots , \qquad \text{(A5.20)}$$

leading to Eq. (2.38).

Similarly, with the aid of Eq. (A1.6), we have

$$\mathrm{Im}\ \varepsilon(\vec{k}, \omega) = -\frac{\pi\omega_p^2}{k^2} \int d\vec{v}\, \delta(\omega - \vec{k} \cdot \vec{v}) \vec{k} \cdot \frac{\partial f(\vec{v})}{\partial \vec{v}} \ . \qquad \text{(A5.21)}$$

Since the decay rate $\gamma_k$ of the plasma oscillation is determined from the solution to $\mathrm{Im}\ \varepsilon(\vec{k}, \omega_k - i\gamma_k) = 0$, we find

$$\gamma_k = \frac{\omega_k}{2} \mathrm{Im}\ \varepsilon(\vec{k}, \omega_k) \ , \qquad \text{(A5.22)}$$

which leads to Eq. (2.41).

When the plasma under consideration is in thermodynamic equilibrium, the velocity distribution is given by the Maxwellian (1.2); Eq. (A5.19) is then evaluated as

$$\varepsilon(\vec{k}, \omega) = 1 + \frac{k_D^2}{k^2} W\left(\frac{\omega}{k(k_B T/m)^{1/2}}\right) \qquad \text{(A5.23)}$$

where we have introduced the **W function** (see e.g., Ichimaru

1973) according to

$$W(Z) = (2\pi)^{-1/2} \int_{-\infty}^{\infty} dx \frac{x}{x - Z - i\eta} \exp(- \frac{x^2}{2}) \quad . \tag{A5.24}$$

This is an analytic function without a singularity in the upper half of the complex Z plane; it is then continued analytically into the lower half-plane. We thus find

$$W(Z) = 1 - Z\exp(- \frac{Z^2}{2}) \int_0^Z dy \exp(\frac{y^2}{2}) + i(\frac{\pi}{2})^{1/2} Z \exp(- \frac{Z^2}{2}) \quad . \tag{A5.25}$$

For the purpose of mathematical analysis, it is useful to derive series expansions of W(Z) for small and large arguments. For $|Z| < 1$, W(Z) can be expressed in a convergent series

$$W(Z) = i(\frac{\pi}{2})^{1/2} Z \exp(- \frac{Z^2}{2}) + 1 - Z^2 + \frac{Z^4}{3} - \dots + \frac{(-1)^{n+1} Z^{2n+2}}{(2n+1)!!} + \dots \tag{A5.26}$$

where

$$(2n + 1)!! = (2n + 1)(2n - 1) \dots 3 \cdot 1 \quad . \tag{A5.27}$$

For large Z, we have an asymptotic series

$$W(Z) = i\left(\frac{\pi}{2}\right)^{1/2} Z \exp\left(-\frac{Z^2}{2}\right)$$

$$-\frac{1}{Z^2} - \frac{3}{Z^4} - \cdots - \frac{(2n-1)!!}{Z^{2n}} - \cdots \quad . \qquad \text{(A5.28)}$$

The static dielectric function is obtained by substituting Eq. (A5.26) in Eq. (A5.23), so that

$$\varepsilon(\vec{k}, 0) = 1 + \frac{k_D^{\,2}}{k^2} \quad , \qquad \text{(A5.29)}$$

the Debye-Hückel screening function. The dispersion relation, Eqs. (2.39) and (2.43), may be derived from the solution to Eq. (6.15) where Eq. (A5.28) is substituted in Eq. (A5.23).

Finally we remark on certain analytic properties of the dielectric function. As we noted in connection with Eq. (6.8) in Section 6.2, the charge density response in the plasma has been calculated in such a way as to follow the application of an external disturbance, that is, in such a way as to satisfy the requirement for the principle of causality. As a consequence, the response function, $[\varepsilon(k, \omega)^{-1} - 1]$, is an analytic function having no singularities in the upper half of the complex $\omega$ plane (Im $\omega > 0$). Choosing a closed contour of integration as shown in Fig. A5.1, one thus finds from Cauchy's theorem

$$\int_{C_1 + C_2} d\omega' \, \frac{1}{\omega' - \omega + i\eta} \left[\frac{1}{\varepsilon(\vec{k}, \omega')} - 1\right] = 0 \quad . \qquad \text{(A5.30)}$$

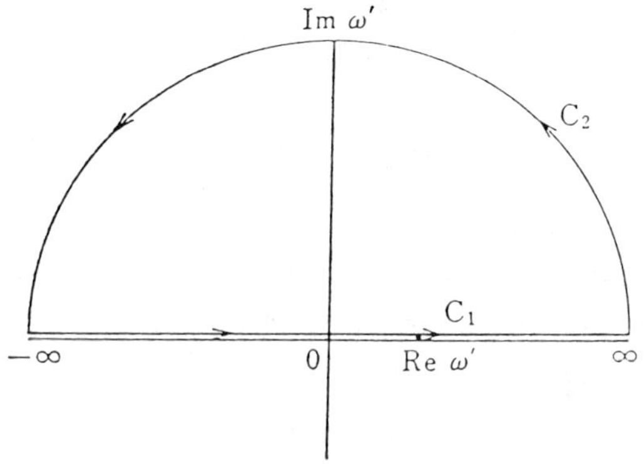


**Figure A5.1.** Contour of integration in Eq. (A5.30).

On the other hand, since

$$\lim_{\omega\to\infty} \left[\frac{1}{\varepsilon(\vec{k},\ \omega)} - 1\right] = 0 \quad , \tag{A5.31}$$

the integration along the semi-infinite contour $C_2$ vanishes. Hence one obtains

$$\int_{-\infty}^{\infty} d\omega' \, \frac{1}{\omega' - \omega + i\eta} \left[\frac{1}{\varepsilon(\vec{k},\ \omega')} - 1\right] = 0 \tag{A5.32}$$

from (A5.30).

With the aid of the relation (A1.6), this integration can be rewritten as

$$\lim_{\eta \to 0} \left[ \int_{-\infty}^{\omega-\eta} + \int_{\omega+\eta}^{\infty} \right] d\omega' \frac{1}{\omega' - \omega} \left[ \frac{1}{\varepsilon(\vec{k}, \omega')} - 1 \right]$$

$$- i\pi \left[ \frac{1}{\varepsilon(\vec{k}, \omega)} - 1 \right] = 0 \quad . \tag{A5.33}$$

The first half of this formula represents what is called Cauchy's principal-value integral, which we shall denote by a symbol P. Splitting Eq. (A5.33) into the real and imaginary parts, we find

$$\mathrm{Im} \frac{1}{\varepsilon(\vec{k}, \omega)} = \frac{1}{\pi} \int_{-\infty}^{\infty} d\omega' \frac{P}{\omega' - \omega} \left[ 1 - \mathrm{Re} \frac{1}{\varepsilon(\vec{k}, \omega')} \right] \tag{A5.34}$$

$$\mathrm{Re} \frac{1}{\varepsilon(\vec{k}, \omega)} - 1 = \frac{1}{\pi} \int_{-\infty}^{\infty} d\omega' \frac{P}{\omega' - \omega} \mathrm{Im} \frac{1}{\varepsilon(\vec{k}, \omega')} \quad . \tag{A5.35}$$

Those are the **Kramers-Kronig relations**, related directly to the requirement for the causality principle in the linear response function.

## Appendix 6

## Static Structure Factor in the Hartree-Fock Approximation

In terms of the creation and annihilation operators of the electrons introduced in Appendix 5, the static structure factor, Eq. (3.20), is expressed as [cf. Eqs. (A5.6) and (A5.7)]

$$S(\vec{k}) = \frac{1}{n} < \sum_{\substack{\vec{p}\vec{p}' \\ \sigma\sigma'}} c^{+}_{\vec{p}\sigma} c_{\vec{p} + \hbar\vec{k}\sigma} c^{+}_{\vec{p}' + \hbar\vec{k}\sigma'} c_{\vec{p}'\sigma'} >$$

$$= 1 + \frac{1}{n} < \sum_{\substack{\vec{p}\vec{p}' \\ \sigma\sigma'}} c^{+}_{\vec{p}' + \hbar\vec{k}\sigma'} c^{+}_{\vec{p} - \hbar\vec{k}\sigma} c_{\vec{p}\sigma} c_{\vec{p}'\sigma'} > . \qquad (A6.1)$$

In the Hartree-Fock approximation, where the correlations are brought about only through the Pauli principle Eq. (A5.5), we find

$$S_{HF}(\vec{k}) = 1 + (n - 1)\delta_{0\vec{k}} - \frac{1}{n} < \sum_{\substack{\vec{p}\sigma \\ \vec{k}\neq 0}} n_{\vec{p} + \hbar\vec{k}\sigma} n_{\vec{p}\sigma} >$$

$$= \begin{cases} n\delta_{0\vec{k}} + \frac{3}{4}\frac{k}{k_F} - \frac{1}{16}\left(\frac{k}{k_F}\right)^3 & (k \leq 2k_F) \\ 1 & (k > 2k_F) \end{cases} \quad . \tag{A6.2}$$

This is the static structure factor in the Hartree-Fock approximation.

Substitution of Eq. (A6.2) in Eq. (3.22) yields the radial distribution function in the Hartree-Fock approximation,

$$g_{HF}(r) = 1 - \frac{9}{2}\left[\frac{\sin k_F r - k_F r \cos k_F r}{k_F^3 r^3}\right]^2 \quad . \tag{A6.3}$$

This function is plotted in Fig. 6.2.

# Appendix 7

## Fluctuation-Dissipation Theorem and its Application

Consider a situation in which an external disturbance

$$a(\vec{\mathbf{r}}, t) = a \exp[i(\vec{\mathbf{k}} \cdot \vec{\mathbf{r}} - \omega t) + \eta t] + (\text{complex conjugate}) \tag{A7.1}$$

is applied to an otherwise uniform physical system in thermodynamic equilibrium; the disturbance couples with the system via an extra Hamiltonian

$$H_{ext}(t) = - \int d\vec{\mathbf{r}}\, A(\vec{\mathbf{r}}) a(\vec{\mathbf{r}}, t) \tag{A7.2}$$

where A represents a physical quantity of the system. A physical quantity B--it can be the same as A--of the system is then disturbed and thereby creates a deviation from the average, $\delta B(\vec{\mathbf{r}}, t)$. Limiting ourselves only to a consideration of the linear response against a weak disturbance, we may express the deviation as

$$\delta B(\vec{r}, t) = b \exp[i(\vec{k} \cdot \vec{r} - \omega t) + \eta t] + (\text{complex conjugate}) \ . \tag{A7.3}$$

In this case

$$K_{BA}(\vec{k}, \omega) = b/a \tag{A7.4}$$

gives a **linear response function** in its general form.

The physical quantities A and B fluctuate in space and time even for a system in thermodynamic equilibrium. The **correlation function** between them is defined as

$$C_{BA}(\vec{r}, t) = \frac{1}{2} \langle [B(\vec{r}' + \vec{r}, t' + t)A(\vec{r}', t') + A(\vec{r}', t')B(\vec{r}' + \vec{r}, t' + t)] \rangle \tag{A7.5}$$

where < ... > refers to the expectation value in the equilibrium state.

Fourier transformation of this correlation function yields a general **structure factor**,

$$S_{BA}(\vec{k}, \omega) = \frac{1}{2\pi} \int d\vec{r} \int_{-\infty}^{\infty} dt \ C_{BA}(\vec{r}, t) \exp[-i(\vec{k} \cdot \vec{r} - \omega t)] \ . \tag{A7.6}$$

The spectral function $S(\vec{k}, \omega)$ of the density fluctuations defined in Eq. (3.16) corresponds to the case where A = B = (number density) in Eq. (A7.6).

According to the general statistical-mechanical treatment (Callan and Welton 1951; Kubo 1957), the linear response function (A7.4) and the structure factor (A7.6) are connected to each other via the relation

$$S_{BA}(\vec{k}, \omega) = -\frac{i\hbar}{4\pi}\coth\left(\frac{\hbar\omega}{2k_BT}\right)[K_{BA}(\vec{k}, \omega) - K_{AB}(-\vec{k}, -\omega)] \ . \tag{A7.7}$$

This is the **fluctuation-dissipation theorem.**

In the limit of $T \to 0$ as in the case of Chapter 6, Eq. (A7.7) becomes

$$S_{BA}(\vec{k}, \omega) = -\frac{i\hbar}{4\pi}[K_{BA}(\vec{k}, \omega) - K_{AB}(-\vec{k}, -\omega)] \ . \tag{A7.8}$$

In the classical limit $\hbar \to 0$, on the other hand, Eq. (A7.7) takes the form

$$S_{BA}(\vec{k}, \omega) = -\frac{ik_BT}{2\pi\omega}[K_{BA}(\vec{k}, \omega) - K_{AB}(-\vec{k}, -\omega)] \ . \tag{A7.9}$$

We next apply the fluctuation-dissipation theorem and derive a relation between the dynamic structure factor (3.16) and the dielectric response function (6.10).

The external potential field, Eq. (6.8), is coupled with the density fluctuation $\rho(\vec{r})$ in the plasma according to

$$H_{ext}(t) = Ze\int d\vec{r}\, \rho(\vec{r})\phi_{ext}(\vec{r}, t) \ . \tag{A7.10}$$

Upon comparison between Eqs. (A7.10) and (A7.2), we find the linear response function (A7.4) appropriate to the calculation of $S(\vec{k}, \omega)$ as

$$K_{\rho\rho}(\vec{k}, \omega) = -\frac{\langle\rho_{ind}\rangle_{\vec{k}\omega}}{Ze(\phi_{ext})_{\vec{k}\omega}} \ . \tag{A7.11}$$

Substitution of Eqs. (6.11) and (6.12) in this formula yields

$$K_{\rho\rho}(\vec{k}, \omega) = \frac{k^2}{4\pi(Ze)^2}\left[1 - \frac{1}{\varepsilon(\vec{k}, \omega)}\right] . \tag{A7.12}$$

As may be clear from the examples of Eqs. (A5.15) and (A5.19), the response functions have symmetries

$$\frac{1}{\varepsilon(\vec{k}, \omega)} - \frac{1}{\varepsilon(-\vec{k}, -\omega)} = 2i \ \mathrm{Im} \frac{1}{\varepsilon(\vec{k}, \omega)} , \tag{A7.13}$$

so that we derive a relation

$$S(\vec{k}, \omega) = - \frac{\hbar k^2}{8\pi^2(Ze)^2} \coth\left(\frac{\hbar\omega}{2k_BT}\right) \mathrm{Im} \frac{1}{\varepsilon(\vec{k}, \omega)} \tag{A7.14}$$

with the aid of Eqs. (A7.7) and (A7.12).

For the electron plasma in metal (Chapter 6) one can take the limit $T \to 0$, so that $(Z = -1)$

$$S(\vec{k}, \omega) = - \frac{\hbar k^2}{4\pi^2 e^2} \mathrm{Im} \frac{1}{\varepsilon(\vec{k}, \omega)} \qquad (\omega \geq 0) . \tag{A7.15}$$

In the derivation of Eq. (A7.15) from Eq. (A7.14), we have taken the limit of $T \to 0$ and then added the values of Eq (A7.14) in the domain $\omega < 0$ to those in the domain $\omega > 0$. Hence Eq. (A7.15) is defined only in the domain $\omega \geq 0$; for $\omega < 0$ it vanishes.

For a classical plasma, since $\hbar|\omega| \ll k_BT$, we obtain

$$S(\vec{k}, \omega) = - \frac{k_BTk^2}{4\pi^2(Ze)^2\omega} \mathrm{Im} \frac{1}{\varepsilon(\vec{k}, \omega)} . \tag{A7.16}$$

This formula is defined over the domain, $-\infty < \omega < \infty$, where we note $S(\vec{k}, \omega) = S(\vec{k}, -\omega)$.

To calculate the static structure factor $S(\vec{k})$ of a classical plasma from Eq. (A7.16) actually in accord with Eq. (3.19), it is convenient to use the Kramers-Kronig relations discussed in Appendix 5. We substitute Eq. (A7.16) in Eq. (3.19) and carry out the $\omega$ integration, which is the same as putting $\omega = 0$ in Eq. (A5.35); hence we find

$$S(\vec{k}) = \frac{k^2}{k_D^2}\left[1 - \frac{1}{\varepsilon(\vec{k}, 0)}\right] . \tag{A7.17}$$

In the RPA, Eq. (A5.29) is applicable, so that Eq. (A7.17) leads to Eq. (4.14).

For a two-component electron-ion plasma in the RPA the dielectric function is expressed as

$$\varepsilon(\vec{k}, \omega) = 1 - \frac{4\pi e^2}{k^2}\chi_0^{(e)}(\vec{k}, \omega) - \frac{4\pi(Ze)^2}{k^2}\chi_0^{(i)}(\vec{k}, \omega) . \tag{A7.18}$$

The dynamic structure factor for the electrons in the classical limit is then given by

$$S(\vec{k}, \omega) = -\frac{k_B T k^2}{4\pi^2 e^2 \omega}\,\mathrm{Im}\left\{\frac{4\pi e^2}{k^2}\chi_0^{(e)}(\vec{k}, \omega) \times \frac{1 - [4\pi(Ze)^2/k^2]\chi_0^{(i)}(\vec{k}, \omega)}{\varepsilon(\vec{k}, \omega)}\right\} . \tag{A7.19}$$

The use of the Kramers-Kronig relations again yields the static structure factor, Eq. (4.17).

## Appendix 8

## Superposition of Dressed Particles

In this appendix we consider a theoretical analysis of the correlation functions and the spectra of fluctuations in a nonequilibrium, uniform and stationary plasma within the random-phase approximation (RPA). We begin by defining the correlation function between the quantities, A and B, and the associated spectral function $S_{AB}(\vec{k}, \omega)$:

$$C_{AB}(\vec{r} - \vec{r}', t - t') \equiv \langle A(\vec{r}, t)B(\vec{r}', t)\rangle$$

$$= \frac{1}{V} \sum_{\vec{k}} \int d\omega \; S_{AB}(\vec{k}, \omega) \exp[i\vec{k} \cdot (\vec{r} - \vec{r}') - i\omega(t - t')] \quad . \tag{A8.1}$$

This correlation function is basically the same as that defined by Eq. (A7.5). The spectral function can be calculated in terms of the Fourier components of A and B as

$$\langle \tilde{A}(\vec{k}, \omega)\tilde{B}(-\vec{k}, \omega')\rangle = \frac{(2\pi)^2}{V} S_{AB}(\vec{k}, \omega)\delta(\omega + \omega') \quad . \tag{A8.2}$$

To calculate the spectral functions of the electromagnetic fluctuations, we note Eq. (A4.3) describing the fluctuations in the microscopic particle distribution. As remarked in connection with Eq. (A4.7), the third, nonlinear term on the right-hand side of Eq. (A4.3) is negligible in the RPA. We then calculate the Fourier components of the spontaneous current density according to

$$\delta\vec{\mathbf{J}}_0(\vec{\mathbf{k}}, \omega) \equiv \sum_\sigma e_\sigma \int d\vec{\mathbf{p}}\ \vec{\mathbf{v}}\delta\tilde{N}_\sigma^{(0)}(\vec{\mathbf{k}}, \omega; \vec{\mathbf{p}})$$

$$= \frac{i\omega}{4\pi}[\overleftrightarrow{\varepsilon}(\vec{\mathbf{k}}, \omega) - (\frac{ck}{\omega})^2 \overleftrightarrow{I}_T] \cdot \delta\vec{\mathbf{E}}(\vec{\mathbf{k}}, \omega) \quad , \tag{A8.3}$$

where

$$\overleftrightarrow{I}_T = \overleftrightarrow{I} - \frac{\vec{\mathbf{k}}\vec{\mathbf{k}}}{k^2} \tag{A8.4}$$

is the transverse projection tensor with respect to $\vec{\mathbf{k}}$. In the derivation of Eq. (A8.3), we have used Eqs. (A3.6) for fluctuations, Eqs. (A4.6) and (A4.9).

We now introduce the **dispersion tensor** $\overleftrightarrow{D}(\vec{\mathbf{k}}, \omega)$ calcualted in accord with

$$\overleftrightarrow{D}(\vec{\mathbf{k}}, \omega) \cdot [\overleftrightarrow{\varepsilon}(\vec{\mathbf{k}}, \omega) - (\frac{ck}{\omega})^2 \overleftrightarrow{I}_T] = \overleftrightarrow{I} \quad . \tag{A8.5}$$

Clearly the dispersion tensor has singularities at those combinations of $\vec{\mathbf{k}}$ and $\omega$ which satisfy the determinant equation

$$\det|\overleftrightarrow{\varepsilon}(\vec{\mathbf{k}}, \omega) - (\frac{ck}{\omega})^2 \overleftrightarrow{I}_T| = 0 \quad . \tag{A8.6}$$

Put another way, Eq. (A8.6) determines the dispersion relations for the propagation of the characteristic electromagnetic fluctuations in the plasma.

Contraction between $\overleftrightarrow{D}(\mathbf{k}, \omega)$ and Eq. (A8.3) yields a solution,

$$\delta\vec{\mathbf{E}}(\vec{\mathbf{k}}, \omega) = \frac{4\pi}{i\omega} \overleftrightarrow{D}(\vec{\mathbf{k}}, \omega) \cdot \delta\vec{\mathbf{J}}_0(\vec{\mathbf{k}}, \omega) \quad . \tag{A8.7}$$

Hence we find the spectral tensor of the electric field fluctuations,

$$\overleftrightarrow{S}_{EE}(\vec{\mathbf{k}}, \omega) = \frac{16\pi^2}{\omega^2} \overleftrightarrow{D}(\vec{\mathbf{k}}, \omega) \cdot \overleftrightarrow{S}_{J_0J_0}(\vec{\mathbf{k}}, \omega) \cdot \overleftrightarrow{D}^{+}(\vec{\mathbf{k}}, \omega) \quad . \tag{A8.8}$$

Here $\overleftrightarrow{D}^{+}(\vec{\mathbf{k}}, \omega)$ is the Hermitian conjugate tensor to $\overleftrightarrow{D}(\vec{\mathbf{k}}, \omega)$, and $\overleftrightarrow{S}_{J_0J_0}(\vec{\mathbf{k}}, \omega)$ refers to the spectral tensor of the electric current-density fluctuations defined by the first equation of Eq. (A8.3).

Explicit calculations of $\overleftrightarrow{S}_{J_0J_0}(\vec{\mathbf{k}}, \omega)$ are straightforward once we recall that $\delta\tilde{N}_\sigma^{(0)}(\vec{\mathbf{k}}, \omega; p)$ are the Fourier components of Eq. (A3.22) representing the fluctuations in an "ideal-gas" system of discrete particles; $X_j^{(0)}(t, t_0)$ refers to the unperturbed phase-space trajectory stemming from the solution to Eqs. (A3.20). For $\langle\vec{\mathbf{E}}\rangle = 0$ and $\langle\vec{\mathbf{B}}\rangle = 0$, we thus find

$$\overleftrightarrow{S}_{J_0J_0}(\vec{\mathbf{k}}, \omega) = \sum_\sigma e_\sigma^2 \int d\vec{\mathbf{p}}\ \vec{\mathbf{v}}\vec{\mathbf{v}} F_\sigma(\vec{\mathbf{p}})\delta(\omega - \vec{\mathbf{k}} \cdot \vec{\mathbf{v}}) \quad . \tag{A8.9}$$

For $\langle\vec{\mathbf{E}}\rangle = 0$ and $\vec{\mathbf{B}} = B\hat{\mathbf{z}}$, we have

$$\overleftrightarrow{S}_{J_0J_0}(\vec{k}, \omega) = \sum_\sigma e_\sigma^2 \sum_{\nu=-\infty}^{\infty} \int d\vec{p}\, \overleftrightarrow{\Pi}_\sigma(p_\perp, p_\parallel; \nu) F_\sigma(\vec{p})\delta(\nu\Omega_\sigma + k_\parallel v_\parallel - \omega) \tag{A8.10}$$

where the notation of Eqs. (A4.12)-(A4.15) has been used.

It is in the form of Eq. (A8.8) that we interpret the spectral tensor of the electric field fluctuations in the RPA as representing the superposition principle of the fields of dynamically screened or dressed test particles without further consideration of the correlation among them. The theoretical scheme for calculation of the fluctuation spectrum through this superposition principle has been extensively investigated (Thompson and Hubbard 1960; Hubbard 1961; Rostoker 1964a, b; Ichimaru 1973); the method is applicable to a stable plasma in a stationary state.

## Appendix 9

## Incoherent Scattering of Electromagnetic Radiation

The differential cross section, Eq. (3.14), for scattering of electromagnetic radiation is calculated through consideration of nonlinear coupling processes as depicted in Fig. A9.1. Those processes are described by the last nonlinear term of Eq. (A4.3), which we shall denote as $\delta\tilde{N}^{NL}(\vec{k}, \omega; \vec{p})$ for the electrons.

We take the incident electromagnetic wave in the form of Eq. (3.1), where $\vec{E}_1 = \hat{n}_1 E_1$ with $\hat{n}_1$ representing a unit polarization vector. Its Fourier components are

$$\vec{E}_1(\vec{k}, \omega) = \hat{n}_1 E_1 \pi\{\delta(\vec{k}, \vec{k}_1)\delta(\omega - \omega_1) + \delta(\vec{k}, -\vec{k}_1)\delta(\omega + \omega_1)\} \quad , \tag{A9.1}$$

where $\delta(\vec{k}, \vec{k}')$ is the three-dimensional Kronecker's delta.

We assume that the frequencies, $\omega$ and $\omega'$, of the scattered, and the incident electromagnetic waves in the last term of Eq. (A4.3) are much greater than the plasma frequency $\omega_p$ and the cyclotron frequency $|\Omega_e|$ of the electrons. Since the latter frequencies represent the characteristic frequencies of the

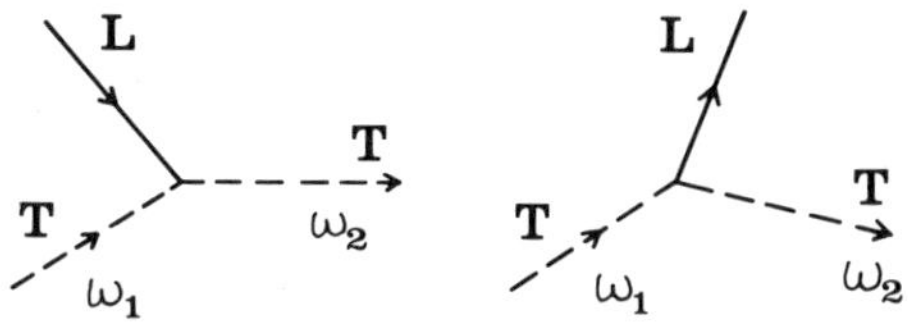


**Figure A9.1.** Nonlinear processes in the incoherent scattering of electromagnetic radiation. T and L designate the transverse and longitudinal excitations, respectively.

electron-density fluctuations $\delta\tilde{N}(\vec{k} - \vec{q}, \omega - \omega'; \vec{p}')$, the foregoing assumption implies

$$|\omega|, |\omega'| \gg |\omega - \omega'| \quad . \tag{A9.2}$$

This also implies that over the characteristic time, $|\omega|^{-1}$ and $|\omega'|^{-1}$, for the change of the transverse fields, the electrons cannot evolve significantly in the phase space, so that we set $\vec{p}' = \vec{p}$. Finally we assume in this appendix that the electrons are nonrelativistic so that $v \ll |\omega'/q| = c$; this observation permits a replacement

$$\overleftrightarrow{T}(\vec{q}, \omega'; \vec{p}) \longrightarrow \overleftrightarrow{I} \tag{A9.3}$$

in the last term of Eq. (A4.3).

We thus find after those simplifications

$$\delta\tilde{N}^{NL}(\vec{k}, \omega; \vec{p}) = \frac{eE_1}{2i\omega V} \hat{n}_1 \cdot \frac{\partial}{\partial\vec{p}}[\delta\tilde{N}(\vec{k} - \vec{k}_1, \omega - \omega_1; \vec{p})$$

$$+ \delta\tilde{N}(\vec{k} + \vec{k}_1, \omega + \omega_1; \vec{p})] \tag{A9.4}$$

where V refers to the volume of the scattering plasma. The electric-current density associated with Eq. (A9.4) is then calculated as

$$\delta\vec{J}^{NL}(\vec{r}, t) = \frac{e^2E_1}{2mV} \hat{n}_1 \sum_{\vec{k}} \int \frac{d\omega}{2\pi} \frac{i}{\omega} [\delta\tilde{n}(\vec{k} - \vec{k}_1, \omega - \omega_1)$$

$$+ \delta\tilde{n}(\vec{k} + \vec{k}_1, \omega + \omega_1)]\exp[i(\vec{k} \cdot \vec{r} - \omega t)] \quad . \tag{A9.5}$$

The radiation field produced by this current-density fluctuation at a distance R from the plasma may be described by a retarded vector potential. Choosing R sufficiently large as compared with the size of the plasma, we write the retarded potential for such scattered waves as

$$\vec{A}(t) = \frac{1}{cR} \int_V d\vec{r}\ \delta\vec{J}^{NL}(\vec{r}, t - \frac{R}{c} + \frac{\hat{k} \cdot \vec{r}}{c}) \tag{A9.6}$$

where $\hat{k}$ is the unit vector in the direction of $\vec{k}$.

Let us now suppose that the observation of the scattered wave is made in a fixed polarization direction $\hat{n}_2$ around the wave vector $\vec{k}_2$. The intensities of the scattered electromagnetic waves are then given by

$$E_2(t) = \hat{n}_2 \cdot \vec{E}_2(t) = \frac{1}{c} \hat{n}_2 \cdot [(\dot{\vec{A}} \times \hat{k}_2) \times \hat{k}_2] \quad , \tag{A9.7a}$$

$$B_2(t) = |\hat{\mathbf{n}}_2 \times \vec{\mathbf{B}}_2(t)| = \frac{1}{c}\, |\hat{\mathbf{n}}_2 \times (\dot{\vec{\mathbf{A}}} \times \hat{\mathbf{k}}_2)| \quad . \tag{A9.7b}$$

The energy flux associated with the scattered wave is calculated from the statistical average < > of Poynting's vector; after a straightforward algebra (e.g., Ichimaru 1973), we find

$$\langle \mathcal{S} \rangle = \frac{c}{4\pi} \langle E_2(t) B_2(t) \rangle$$

$$= \hat{\mathbf{k}}_2 \frac{c E_1^{\,2}}{8\pi} \left(\frac{e^2}{mc^2}\right)^2 \frac{V}{R^2} (\hat{\mathbf{n}}_1 \cdot \hat{\mathbf{n}}_2)^2$$

$$\times \sum_{\vec{\mathbf{k}}} \int_{-\infty}^{\infty} d\omega S(\vec{\mathbf{k}}, \omega) \delta(\vec{\mathbf{k}} + \vec{\mathbf{k}}_1, \frac{\omega + \omega_1}{c} \hat{\mathbf{k}}_2) \tag{A9.8}$$

where $S(\vec{\mathbf{k}}, \omega)$ is the dynamic structure factor of the electrons defined in terms of Eq. (3.12).

In this expression, the quantity $(cE_1^{\,2}/8\pi)$ is the time-averaged energy flux of the incident wave. Hence we obtain from Eq. (A9.8) the differential cross section for the transfer of momentum $\hbar\vec{\mathbf{k}}$ (corresponding to scattering into a solid angle $d\Omega$) and energy $\hbar\omega$ from the plasma to the electromagnetic wave as (e.g., Salpeter 1960; Rosenbluth and Rostoker 1962)

$$\frac{d^2 Q}{d\Omega d\omega} = V\left(\frac{e^2}{mc^2}\right)^2 (\hat{\mathbf{n}}_1 \cdot \hat{\mathbf{n}}_2)^2 S(\vec{\mathbf{k}}, \omega) \tag{A9.9}$$

where $\vec{\mathbf{k}} = \vec{\mathbf{k}}_2 - \vec{\mathbf{k}}_1$ and $\omega = \omega_2 - \omega_1$.

## References


Abe, R., 1959, Prog. Theor. Phys. **21**, 475.

Aigrain, P., 1961, in **Proc. Intern. Conf. on Semicond. Phys.**, Prague, 1960 (Academic, New York), p. 224.

Alastuey, A., and B. Jancovici, 1978, Astrophys. J. **226**, 1034.

Alfvén, H., 1942, Nature **150**, 405.

Alfvén, H., and C. G. Fälthammer, 1963, **Cosmic Electrodynamics** (Oxford Univ. Press, New York).

Allen, T. K., W. R. Baker, R. V. Pyle, and J. M. Wilcox, 1959, Phys. Rev. Lett. **2**, 383.

Baus, M., and J. P. Hansen, 1980, Phys. Rep. **59**, 1.

Berggren, K. F., 1970, Phys. Rev. A **1**, 1783.

Boercker, D. B., F. J. Rogers, and H. E. DeWitt, 1982, Phys. Rev. A **25**, 1623.

Bohm, D., and E. P. Gross, 1949a, Phys. Rev. **75**, 1851.

________, and E. P. Gross, 1949b, Phys. Rev. **75**, 1864.

Bohm, D., and D. Pines, 1951, Phys. Rev. **82**, 625.

________, and D. Pines, 1953, Phys. Rev. **92**, 609.

Bolton, C. T., 1975, Astrophys. J. **200**, 269.

Bowers, R., C. Legendy, and F. Rose, 1961, Phys. Rev. Lett. **7**, 339.

Bowles, K. L., 1958, Phys. Rev. Lett. 1, 454.

______________, 1961, J. Res. N.B.S. **65D**, 1.

Brandt, W., A. Ratkowski, and R. H. Ritchie, 1974, Phys. Rev. Lett. **33**, 1325.

Breuckner, K. A., and S. Jorna, 1974, Rev. Mod. Phys. **46**, 325.

Brush, S. G., H. L. Sahlin, and E. Teller, 1966, J. Chem. Phys. **45**, 2102.

Buchsbaum, S. J., and J. K. Galt, 1961, Phys. Fluids **4**, 1514.

Buneman, O., 1959, Phys. Rev. **115**, 503.

Callen, H. B., and T. A. Welton, 1951, Phys. Rev. **83**, 34.

Ceperley, D. M., and B. J. Alder, 1980, Phys. Rev. Lett. **45**, 566.

Cole, M. W., 1974, Rev. Mod. Phys. **46**, 451.

Davidson, K., and J. P. Ostriker, 1973, Astrophys. J. **179**, 585.

Deacon, D.A.G., L. R. Elias, J.M.J. Madey, G. J. Ramian, H. A. Schwettman, and T. I. Smith, 1977, Phys. Rev. Lett. **38**, 892.

Debye, P., and E. Hückel, 1923, Physik. Z. **24**, 185.

Derfler, H., and T. C. Simonen, 1966, Phys. Rev. Lett. **17**, 172.

DeWitt, H. E., 1976, Phys. Rev. A **14**, 1290.

DeWitt, H. E., H. C. Graboske, and M. S. Cooper, 1973, Astrophys. J. **181**, 439.

DuBois, D. F., 1959, Ann. Phys. (N.Y.) **8**, 24.

Eardley, D. M., and A. P. Lightman, 1975, Astrophys. J. **200**, 187.

Eardley, D. M., A. P. Lightman, and S. L. Shapiro, 1975, Astrophys. J. (Lett.) **199**, L153.

Elias, L. R., W. M. Fairbank, J.M.J. Madey, H. A. Schwettman, and T. I. Smith, 1976, Phys. Rev. Lett. **36**, 717.

Feynman, R. P., 1972, **Statistical Mechanics** (Benjamin, New York).

Flowers, E., and N. Itoh, 1976, Astrophys. J. **206**, 218.

Freidberg, J. P., 1982, Rev. Mod. Phys. **54**, 801.

Galam, S., and J. P. Hansen, 1976, Phys. Rev. A **14**, 816.

Geldart, D.J.W., and S. H. Vosko, 1966, Can. J. Phys. **44**, 2137.

Gell-Mann, M., and K. A. Brueckner, 1957, Phys. Rev. **106**, 364.

Gemmell, D. S., J. Remillieux, J. C. Poizat, M. J. Gaillard, R. E. Holland, and Z. Vager, 1975, Phys. Rev. Lett. **34**, 1420.

Giacconi, R., H. Gursky, E. M. Kellogg, E. Schreier, and H. Tananbaum, 1971, Astrophys. J. (Lett.) **167**, L67.

Gibbons, P. C., S. E. Schnatterly, J. J. Ritsko, and J. R. Fields, 1976, Phys. Rev. B **13**, 2451.

Glicksman, M., 1971, in **Solid State Physics**, edited by H. Ehrenreich, F. Seitz, and D. Turnbull (Academic, New York), Vol. 26, p. 275.

Goldman, M. V., 1984, Rev. Mod. Phys. **56**, 709.

Gordon, W. E., 1958, Proc. IRE **46**, 1824.

Gott, Yu. B., M. S. Ioffe, and V. G. Telkovsky, 1962, Nucl. Fusion Suppl. **3**, 1045.

Grimes, C. C., 1978, Surf. Sci. **73**, 397.

Hansen, J. P., 1973, Phys. Rev. A **8**, 3096.

Hansen, J. P., and I. R. McDonald, 1976, **Theory of Simple Liquids** (Academic, New York).

Hedin, L., and B. I. Lundqvist, 1971, J. Phys. C **4**, 2064.

Heine, V., P. Nozières, and J. W. Wilkins, 1966, Philos. Mag. **13**, 741.

Hewish, A., S. J. Bell, J.D.H. Pilkington, P. F. Scott, and R. A. Collins, 1968, Nature **217**, 709.

Holt, S. S., E. A. Boldt, L. J. Kaluzienskii, and P. J. Serlemitsos, 1975, Nature **256**, 108.

Hopf, F. A., P. Meystre, M. O. Scully, and W. H. Louisell, 1976, Opt. Commun. **18**, 413.

Hubbard, J., 1958, Proc. R. Soc. London, Ser. A **243**, 336.

__________, 1961, Proc. R. Soc. London, Ser. A **260**, 114.

Hubbard, W. B., and M. Lampe, 1969, Astrophys. J. Suppl. **18**, 297.

Ichimaru, S., 1962, Ann. Phys. (N.Y.) **20**, 78.

__________, 1970, Phys. Rev. A **2**, 494.

__________, 1973, **Basic Principles of Plasma Physics** (Benjamin, Reading, Mass.).

__________, 1976, Astrophys. J. **208**, 701.

__________, 1977, Astrophys. J. **214**, 840.

____________, 1982, Rev. Mod. Phys. **54**, 1017.

Ichimaru, S., and N. Iwamoto, 1978, J. Phys. Soc. Jpn. **44**, 1004.

Ichimaru, S., and H. Iyetomi, 1983, J. Phys. Soc. Jpn. **52**, 1730.

Ichimaru, S., and S. Tanaka, 1985, Phys. Rev. A **32**, No. 3.

Ichimaru, S., and K. Utsumi, 1981, Phys. Rev. B **24**, 7385.

Ikezi, H., 1973, Phys. Fluids **16**, 1668.

Ioffe, M. S., 1965, in **Plasma Physics** (IAEA, Vienna), p. 421.

Isihara, A., and E. W. Montroll, 1971, Proc. Nat. Acad. Sci. USA **68**, 3111.

Itoh, N., S. Ichimaru, and S. Nagano, 1978, Phys. Rev. B **17**, 2862.

Itoh, N., S. Mitake, H. Iyetomi, and S. Ichimaru, 1983, Astrophys. J. **273**, 774.

Itoh, N., H. Totsuji, and S. Ichimaru, 1977, Astrophys. J. **218**, 477 [Erratum: Astrophys. J. **220**, 742 (1978)].

Itoh, N., H. Totsuji, S. Ichimaru, and H. E. DeWitt, 1979, Astrophys. J. **234**, 1079 [Erratum: Astrophys. J. **239**, 414 (1980)].

Ivanov, Yu. V., V. B. Mintsev, V. E. Fortov, and A. N. Dremin, 1976, Zh. Eksp. Teor. Fiz. **71**, 216 [Sov. Phys. JETP **44**, 112 (1976)].

Iyetomi, H., 1984, Prog. Theor. Phys. **71**, 427.

Iyetomi, H., and S. Ichimaru, 1983a, Phys. Rev. A **27**, 1734.

____________, and S. Ichimaru, 1983b, Phys. Rev. A **27**, 3241.

Jephcott, D. F., 1959, Nature **183**, 1652.

Kaneko, S., 1960, J. Phys. Soc. Jpn. **15**, 1685.

Kaner, E. A., and V. G. Skobov, 1966, Usp. Fiz. Nauk **89**, 367 [Sov. Phys. Usp. **9**, 480 (1967)].

Kim, H. C., R. L. Stenzel, and A. Y. Wong, 1974, Phys. Rev. Lett. **33**, 886.

Klimontovich, Yu. L., 1957, Zh. Eksp. Teor. Fiz. **33**, 982 [Sov. Phys. JETP **6**, 753 (1958)].

Konstantinov, O. V., and V. I. Perel', 1960, Zh. Eksp. Teor. Fiz. **38**, 161 [Sov. Phys. JETP **11**, 117 (1960)].

Krall, N. A., and A. S. Trivelpiece, 1973, **Principles of Plasma Physics** (McGraw Hill, New York).

Kubo, R., 1957, J. Phys. Soc. Jpn. **12**, 570.

Kukkonen, C. A., and A. W. Overhauser, 1979, Phys. Rev. B **20**, 550.

Lamb, F. K., C. J. Pethick, and D. Pines, 1973, Astrophys. J. **184**, 271.

Landau, L. D., 1946, J. Phys. (USSR) **10**, 25.

Langmuir, I., 1929, Phys. Rev. **33**, 954.

Lehnert, B., 1954, Phys. Rev. **94**, 815.

Lindhard, J., 1954, K. Dan. Vidensk, Selsk. Mat.-Fys. Medd. (8) **28**, 1.

Longmire, C. L., 1963, **Elementary Plasma Physics** (Interscience, New York).

Lundqvist, S., 1949, Phys. Rev. **76**, 1805.

Malmberg, J. H., and C. B. Wharton, 1964, Phys. Rev. Lett. **13**, 184.

Malmberg, J. H., and C. B. Wharton, 1967, Phys. Rev. Lett. **19**, 775.

Manchester, R. N., and J. H. Taylor, 1977, **Pulsars** (Freeman, San Francisco).

Manchester, R. N., J. H. Taylor, and G. R. Huguenin, 1975, Astrophys. J. **196**, 83.

Metropolis, N., A. W. Rosenbluth, M. N. Rosenbluth, A. H. Teller, and E. Teller, 1953, J. Chem. Phys. **21**, 1087.

Michel, F. C., 1982, Rev. Mod. Phys. **54**, 1.

Minoo, H., C. Deutsch, and J. P. Hansen, 1976, Phys. Rev. A **14**, 840.

Miyamoto, K., 1980, **Plasma Physics for Nuclear Fusion** (The MIT Press, Cambridge, Mass.).

Morita, T., 1960, Prog. Theor. Phys. **23**, 829.

Ng, K. C., 1974, J. Chem. Phys. **61**, 2680.

Nishikawa, K., H. Hojo, K. Mima, and H. Ikeji, 1974, Phys. Rev. Lett. **33**, 148.

Novikov, I. D., and K. S. Thorne, 1973, in **Black Holes**, Les Houches 1972, edited by C. DeWitt and B. DeWitt (Gordon and Breach, New York), p. 343.

Nozières, P., and D. Pines, 1958, Phys. Rev. **111**, 442.

Oda, M., P. Gorenstein, H. Gursky, E. Kellogg, E. Schreier, H. Tananbaum, and R. Giacconi, 1971, Astrophys. J. (Lett.) **166**, L1.

Ohtsuki, Y. H., 1983, **Charged Beam Interaction with Solids** (Taylor and Francis, London).

O'Neil, T., and N. Rostoker, 1965, Phys. Fluids **8**, 1109.

Pathak, K. N., and P. Vashishta, 1973, Phys. Rev. B **7**, 3649.

Percus, J. K., and G. J. Yevick, 1958, Phys. Rev. **110**, 1.

Pineo, V. C., L. G. Kraft, and H. W. Briscoe, 1960, J. Geophys. Res. **65**, 2629.

Pines, D., 1963, **Elementary Excitation in Solids** (Benjamin, New York).

Pines, D., and D. Bohm, 1952, Phys. Rev. **85**, 338.

Pines, D., and P. Nozières, 1966, **The Theory of Quantum Liquids** (Benjamin, New York), Vol. I.

Pringle, J. E., and M. J. Rees, 1972, Astron. Astrophys. **21**, 1.

Puff, R. D., 1965, Phys. Rev. **137**, 406A.

Rappaport, S., R. Doxsey, and W. Zauman, 1971, Astrophys. J. (Lett.) **168**, L43.

Raether, H., 1980, **Excitations of Plasmons and Interband Transitions by Electrons** (Springer, Berlin).

Ribe, F. L., 1975, Rev. Mod. Phys. **47**, 7.

Rice, S. A., and P. Gray, 1965, **The Statistical Mechanics of Simple Liquids** (Interscience, New York).

Rickett, B. J., T. H. Hankins, and J. M. Cordes, 1975, Astrophys. J. **201**, 425.

Roberts, D. H., and P. A. Sturrock, 1973, Astrophys. J. **181**, 161.

Rosenbluth, M. N., and N. Rostoker, 1962, Phys. Fluids **5**, 776.

Rosenfeld, Y., and N. W. Ashcroft, 1979, Phys. Rev. A **20**, 1208.

Rostoker, N., 1964a, Phys. Fluids **7**, 479.

____________, 1964b, Phys. Fluids **7**, 491.

Rothschild, R. E., E. A. Boldt, S. S. Holt, and P. J. Serlemitsos, 1974, Astrophys. J. (Lett.) **189**, L13.

Ruderman, M. A., and P. G. Sutherland, 1975, Astrophys. J. **196**, 51.

Salpeter, E. E., 1954, Aust. J. Phys. **7**, 373.

______________, 1960, Phys. Rev. **120**, 1528.

Salpeter, E. E., and H. M. Van Horn, 1969, Astrophys. J. **155**, 183.

Sanford, P. W., J. C. Ives, S. J. Bell Burnell, K. O. Mason, and P. Murdin, 1975, Nature **256**, 109.

Schatzman, E., 1958, **White Dwarfs** (North-Holland, Amsterdam).

Schneider, T., 1971, Physica **52**, 481.

Schreier, E., H. Gursky, E. Kellogg, H. Tananbaum, and R. Giacconi, 1971, Astrophys. J. (Lett.) **170**, L21.

Shakura, N. I., and R. A. Sunyaev, 1973, Astron. Astrophys. **24**, 337.

Shapiro, S. L., A. P. Lightman, and D. M. Eardley, 1976, Astrophys. J. **204**, 187.

Shapiro, S. L., and S. A. Teukolsky, 1983, **Black Holes, White Dwarfs, and Neutron Stars** (Wiley, New York).

Singwi, K. S., and M. P. Tosi, 1981, in **Solid State Physics**, edited by H. Ehrenreich, F. Seitz, and D. Turnbull (Academic, New York), Vol. 36, p. 177.

Singwi, K. S., M. P. Tosi, R. H. Land, and A. Sjölander, 1968, Phys. Rev. **176**, 589.

Slattery, W. L., G. D. Doolen, and H. E. DeWitt, 1980, Phys. Rev. A **21**, 2087.

______________, G. D. Doolen, and H. E. DeWitt, 1982, Phys. Rev. A **26**, 2255.

Spitzer, Jr., L., 1962, **Physics of Fully Ionized Gases**, 2nd ed. (Interscience, New York).

Spitzer, Jr., L., and R. Härm, 1953, Phys. Rev. **89**, 977.

Springer, J. F., M. A. Pokrant, and F. A. Stevens, 1973, J. Chem. Phys. **58**, 4863.

Stevenson, D. J., and N. W. Ashcroft, 1974, Phys. Rev. A **9**, 782.

Sturrock, P. A., 1971, Astrophys. J. **164**, 529.

Tago, K., K. Utsumi, and S. Ichimaru, 1981, Prog. Theor. Phys. **65**, 54.

Tananbaum, H., H. Gursky, E. M. Kellogg, R. Giacconi, and C. Jones, 1972a, Astrophys. J. (Lett.) **177**, L5.

Tananbaum, H., H. Gursky, E. M. Kellogg, R. Levinson, E. Schreier, and R. Giacconi, 1972b, Astrophys. J. (Lett.) **174**, L143.

Tange, T., and S. Ichimaru, 1974, J. Phys. Soc. Jpn. **36**, 1437.

Taylor, J. H., R. N. Manchester, and G. R. Huguenin, 1975, Astrophys. J. **195**, 513.

Thompson, W. B., and J. Hubbard, 1960, Rev. Mod. Phys. **32**, 714.

Thorne, K. S., and R. H. Price, 1975, Astrophys. J. (Lett.) **195**, L101.

Toigo, F., and T. O. Woodruff, 1971, Phys. Rev. B **4**, 371.

Tonks, L., and I. Langmuir, 1929a, Phys. Rev. **33**, 195.

__________, and I. Langmuir, 1929b, Phys. Rev. **34**, 876.

Totsuji, H., 1978, Phys. Rev. A **17**, 399.

Totsuji, H., and S. Ichimaru, 1973, Prog. Theor. Phys. **50**, 753.

__________, and S. Ichimaru, 1974, Prog. Theor. Phys. **52**, 42.

Utsumi, K., and S. Ichimaru, 1980a, Phys. Rev. B **22**, 1522.

__________, and S. Ichimaru, 1980b, Phys. Rev. B **22**, 5203.

__________, and S. Ichimaru, 1981, Phys. Rev. B **23**, 3291.

__________, and S. Ichimaru, 1982, Phys. Rev. A **26**, 603.

Vager, Z., and D. S. Gemmell, 1976, Phys. Rev. Lett. **37**, 1352.

Van Horn, H. M., 1971, in **White Dwarfs**, edited by W. J. Luyten (Reidel, Dordrecht), p. 136.

Van Horn, H. M., and E. E. Salpeter, 1967, Phys. Rev. **157**, 751.

Van Hoven, G., 1966, Phys. Rev. Lett. **17**, 169.

van Leeuwen, J.M.J., J. Groeneveld, and J. De Boer, 1959, Physica **25**, 792.

Vashishta, P., and K. S. Singwi, 1972, Phys. Rev. B **6**, 875.

Vosko, S. H., L. Wilk, and M. Nusair, 1980, Can. J. Phys. **58**, 1200.

Washimi, H., and T. Taniuti, 1966, Phys. Rev. Lett. **17**, 996.

Weinberg, S., 1972, **Gravitation and Cosmology** (Wiley, New York).

Wigner, E. P., 1934, Phys. Rev. **46**, 1002.

_____________, 1938, Trans. Faraday Soc. **34**, 678.

Wilcox, J. M., F. I. Boley, and A. W. DeSilva, 1960, Phys. Fluids **3**, 15.

Williams, G. A., and G. E. Smith, 1964, IBM Journal **8**, 276.

Williams, R. H., and H. E. DeWitt, 1969, Phys. Fluids **12**, 2326.

Wong, A. Y., and P. Y. Cheung, 1984, Phys. Rev. Lett. **52**, 1222.

Zakharov, V. E., 1967, Zh. Eksp. Teor. Fiz. **53**, 1745 [Sov. Phys. JETP **26**, 994 (1968)].

_____________, 1972, Zh. Eksp. Teor. Fiz. **62**, 1745 [Sov. Phys. JETP **35**, 908 (1972)].

Ziman, J. M., 1961, Philos. Mag. **6**, 1013.

Index